# Introduction

Dealing with numbers is a very essential and important part of our life. Numbers are all around us everywhere we go. They come only second to language of priorities of what to know and master. In many cases numbers and math even cross the boundaries of language and culture. They are simply more universal then language.

Despite the glory of numbers and numerical operations, we find that so many people shy away from dealing with numbers and math. Sometimes that is because numbers and math are often dry and sharp edged. In our continued effort, we introduce a brand new numerical puzzle to help people play with numbers in a fun, smart, brain-enhancing way. In this book, you will see for yourself that JUGGLE 3 is a fantastic way for the old and the young to practice thinking and problem solving techniques in a numerical way.

There are more than 400 problems/puzzles in this book. In each one of them, there are 3 missing numbers and just enough information to find them. The missing three numbers are connected to each other and to the given information using one of the 3 basic numerical operations namely ADDITION, SUBTRACTION, and MULTIPLICATION. ADDITION puzzles in this book outnumber the other two operations puzzles.

This book will be a great addition to classroom as well. It could be used as activity/workbook for any age higher than 6 years. It could also be used for timed contests.

Juggle 3

## Instructions

In each of the given puzzles find 3 numbers to put in the empty circles such that they fulfill the following easy criteria.

In ADDITION puzzles, every two adjacent numbers you insert in the empty circles, should add to the number in the middle between them. The solution should go around consistently closing the loop.

In SUBTRACTION puzzles, every two adjacent numbers you insert in the empty circles, should subtract (the larger number minus the smaller one) to the number in the middle between them. The solution should go around consistently closing the loop. At the same time the 3 numbers solution should add up to the number in the center of the triangle. This condition is needed here to make sure there is only one solution to the puzzle.

In MULTIPLICATION puzzles, every two adjacent numbers you insert in the empty circles, should multiply to the number in the middle between them. The solution should go around consistently closing the loop.

Have fun with a brain workout.

# Juggle 3

By I.M. (Omar) Minyawi

# Juggle 3 - Addition

**1.**

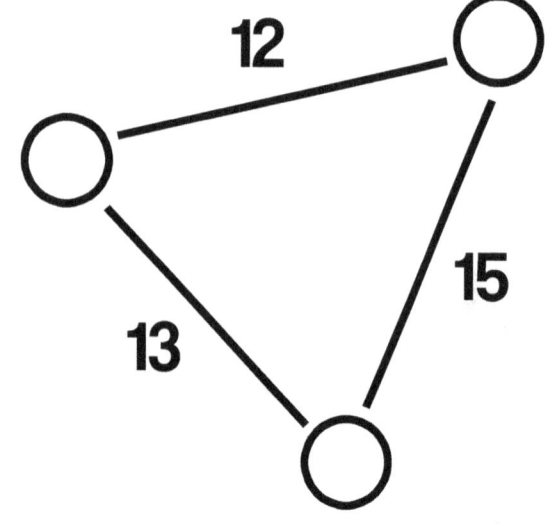

**2.**

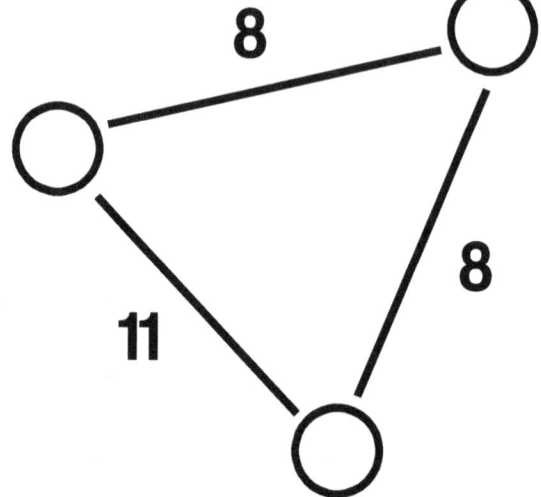

**3.**

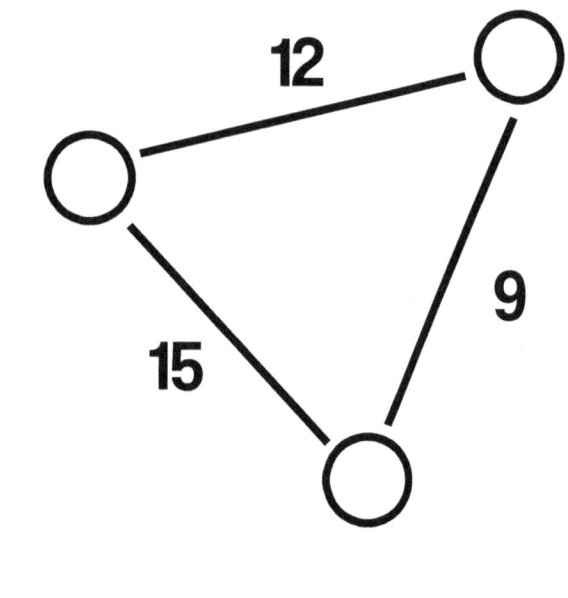

**4.**

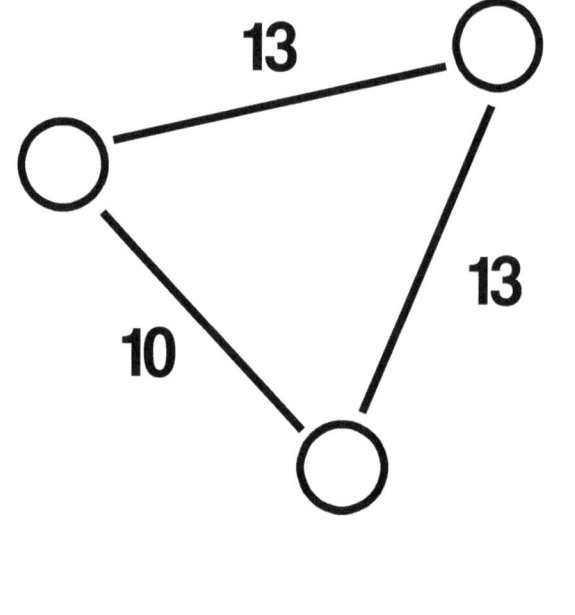

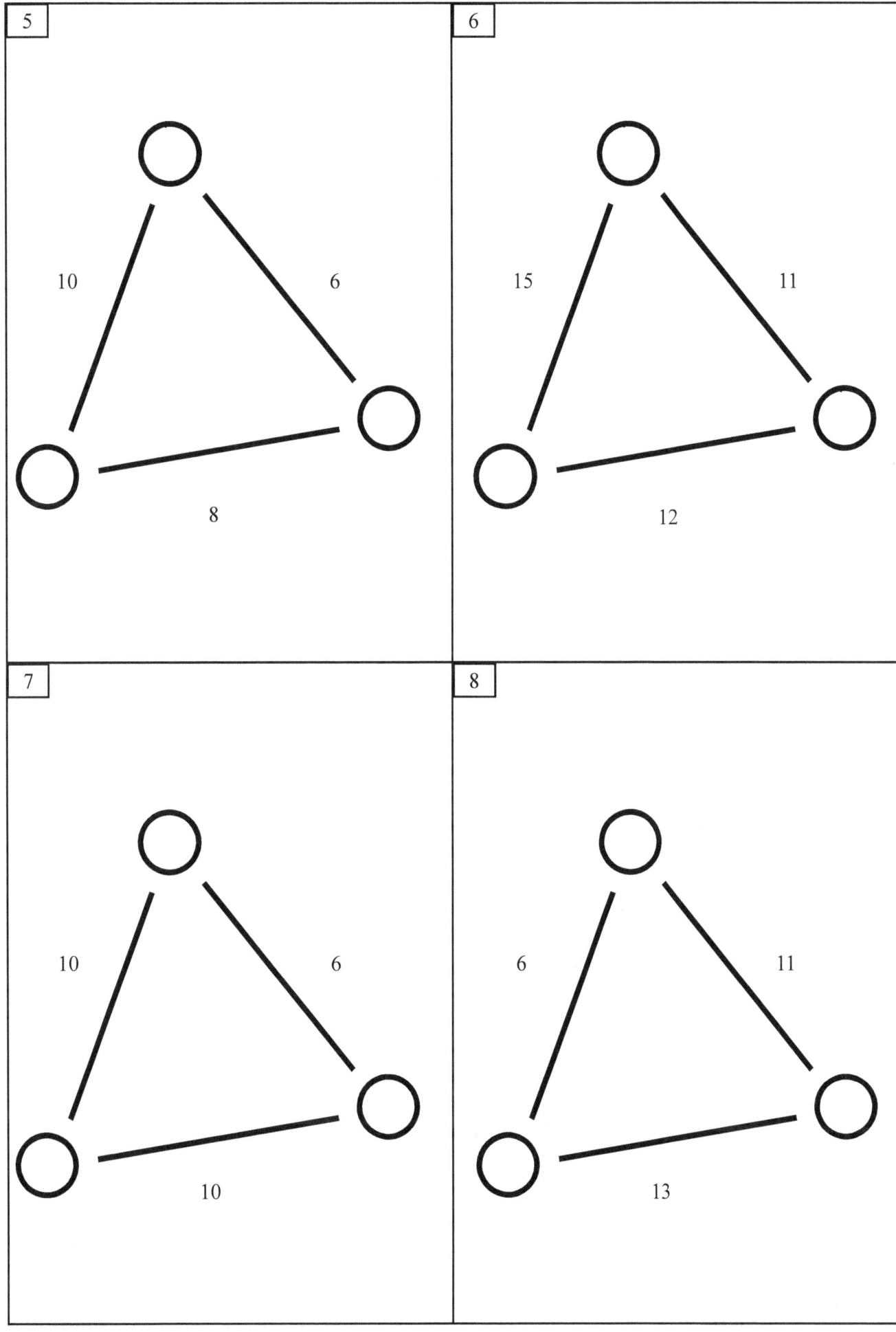

9

```
        O
      /   \
   6 /     \ 4
    /       \
   O         O
      10
```

10

```
        O
      /   \
  10 /     \ 11
    /       \
   O         O
      7
```

11

```
        O
      /   \
  12 /     \ 9
    /       \
   O         O
      5
```

12

```
        O
      /   \
  10 /     \ 6
    /       \
   O         O
      6
```

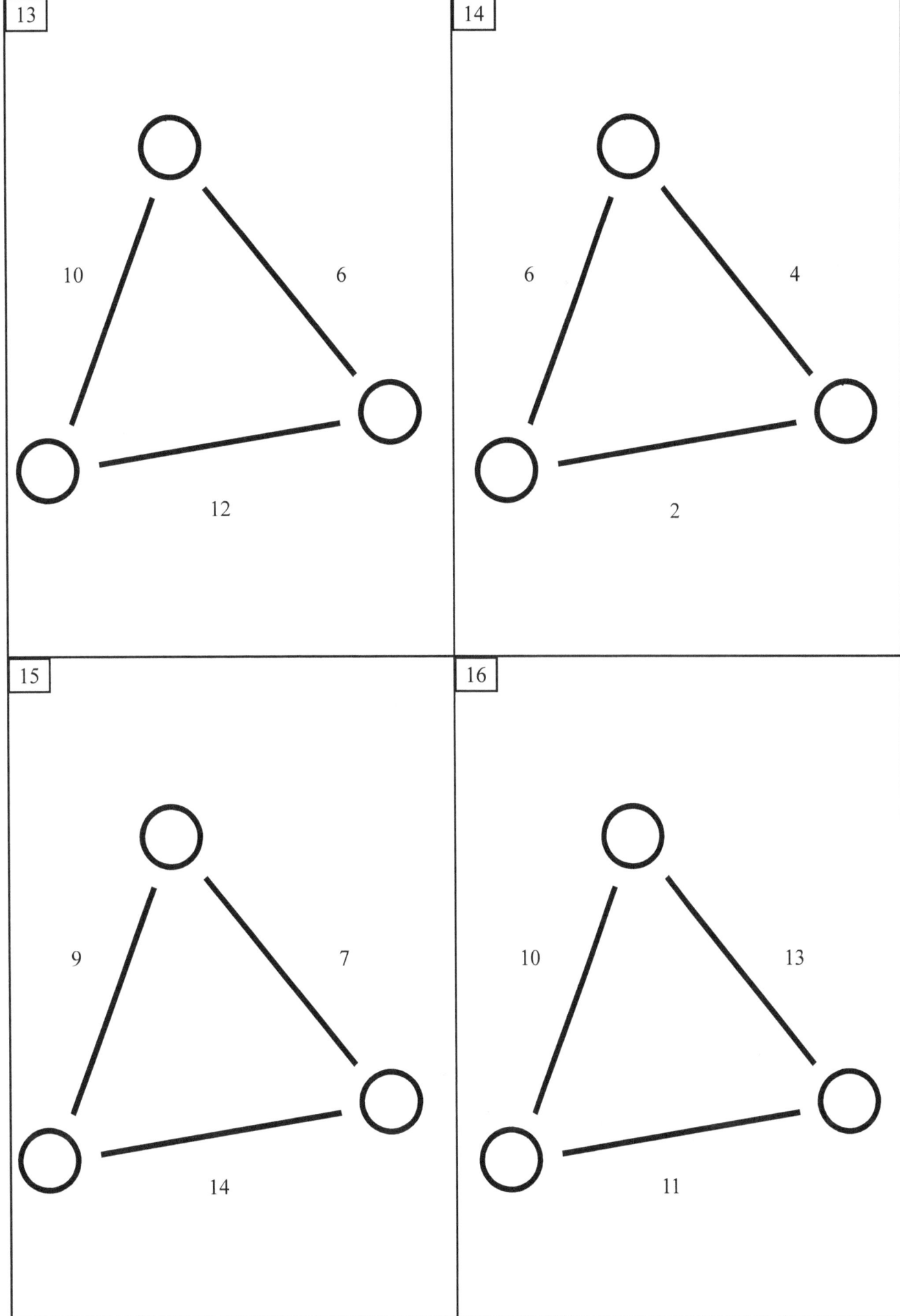

**juggle 3**

17

11    10

5

18

11    8

5

19

13    9

6

20

7    8

15

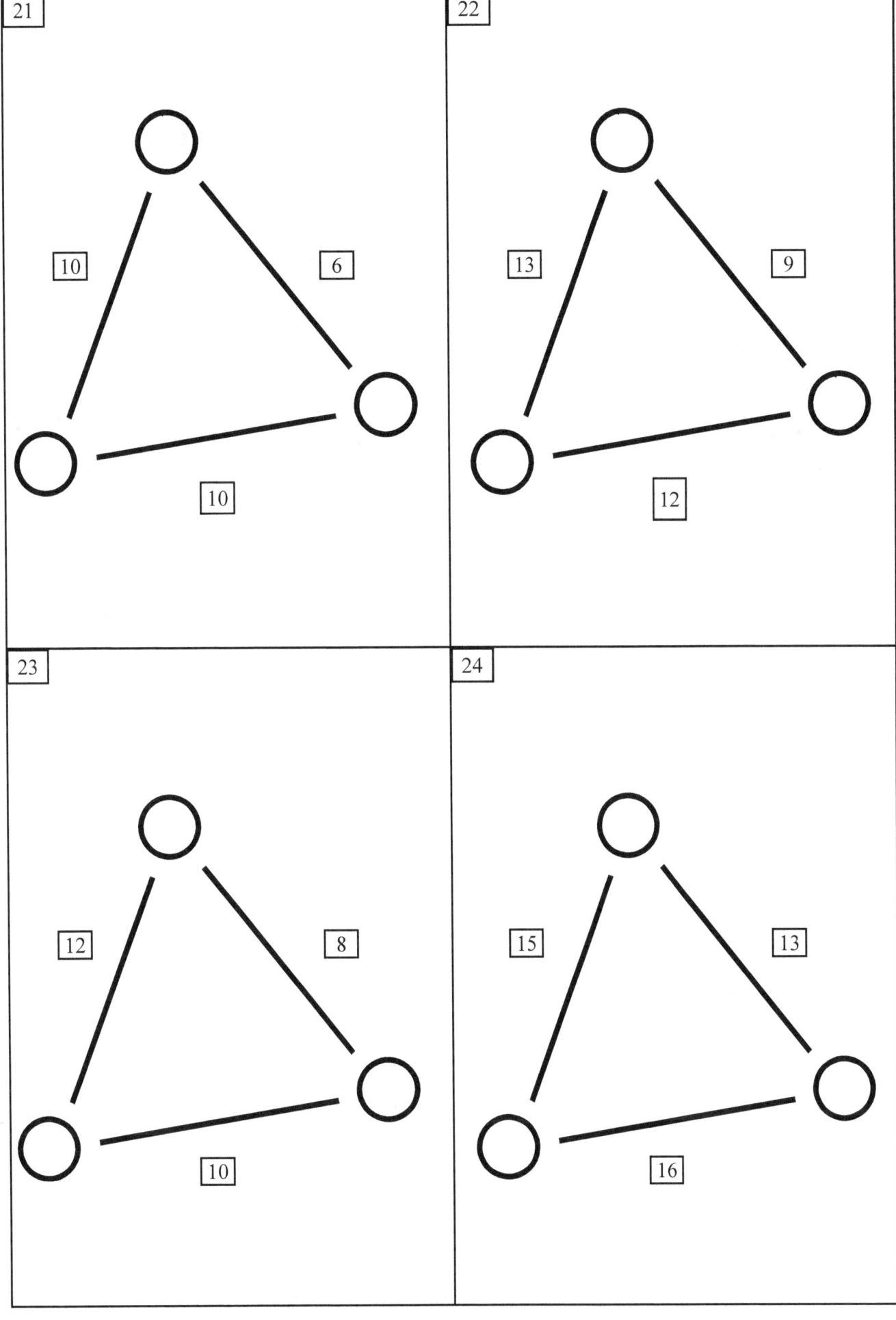

**juggle 3**

25

12

13

11

26

9

10

9

27

12

10

6

28

11

8

7

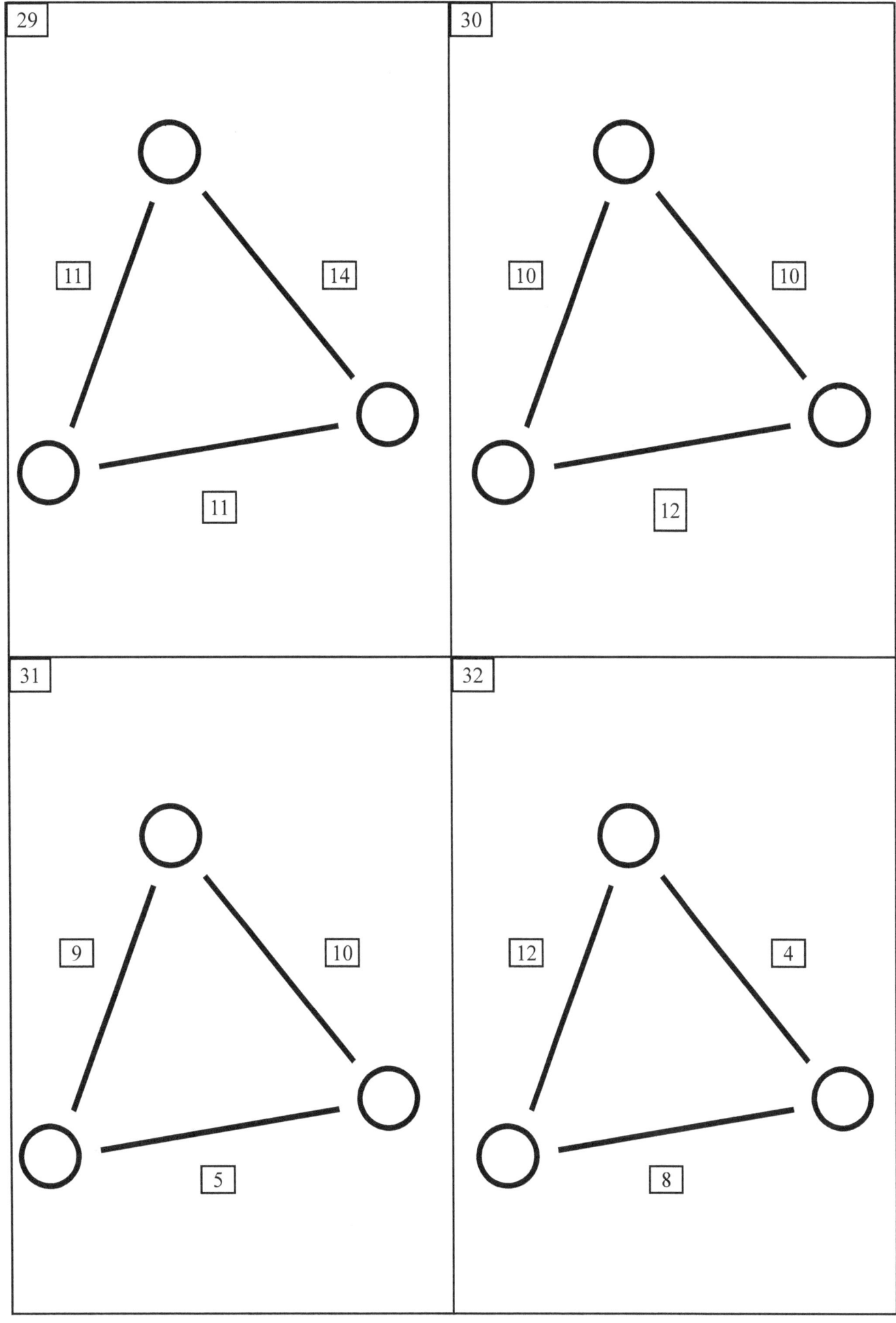

33

8    6

14

34

10    8

12

35

13    12

7

36

12    9

7

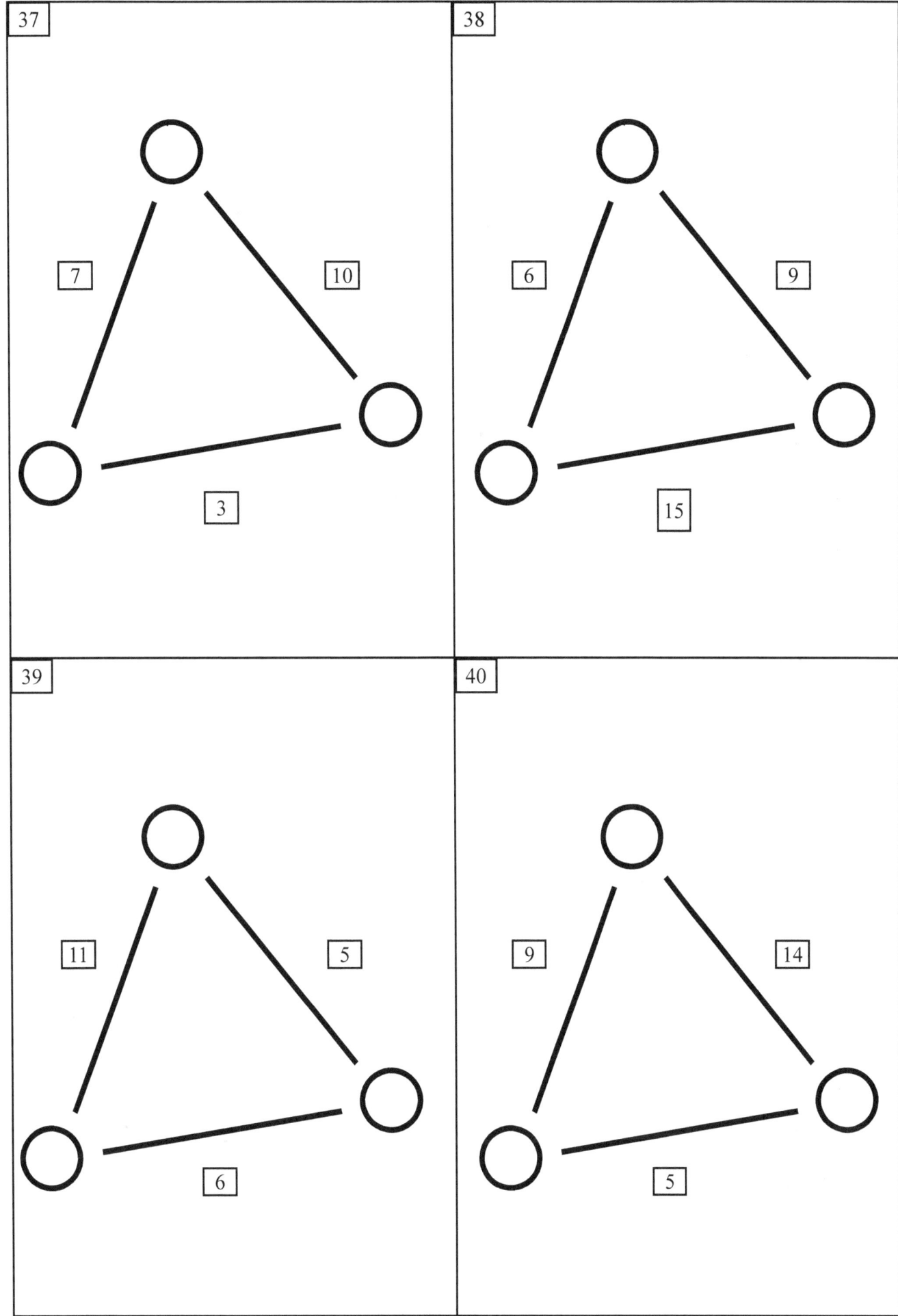

41

10    4

6

42

12    5

9

43

10    6

14

44

9    5

10

**juggle 3**

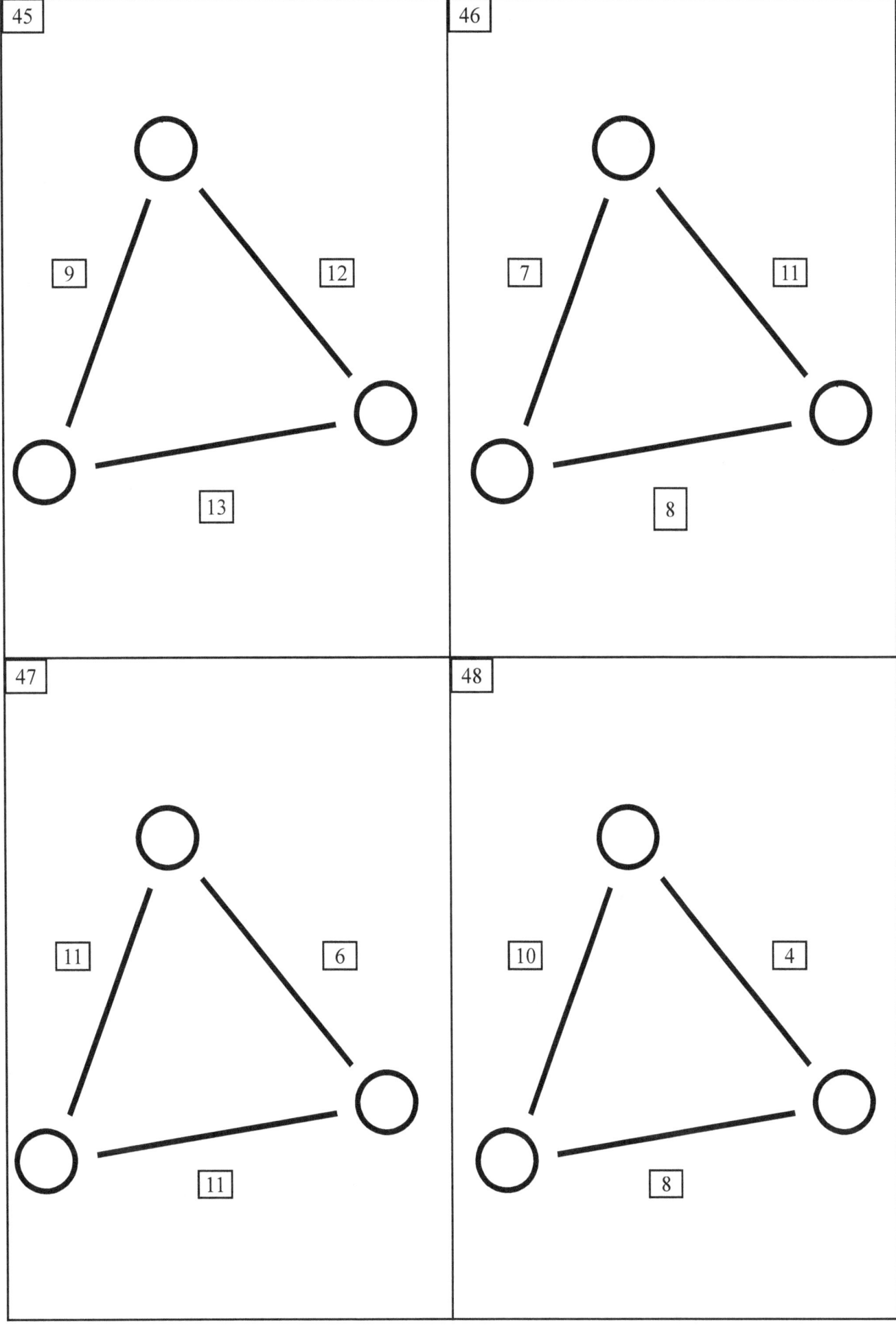

**juggle 3**

49

9   7

6

50

10   6

12

51

9   4

7

52

11   6

13

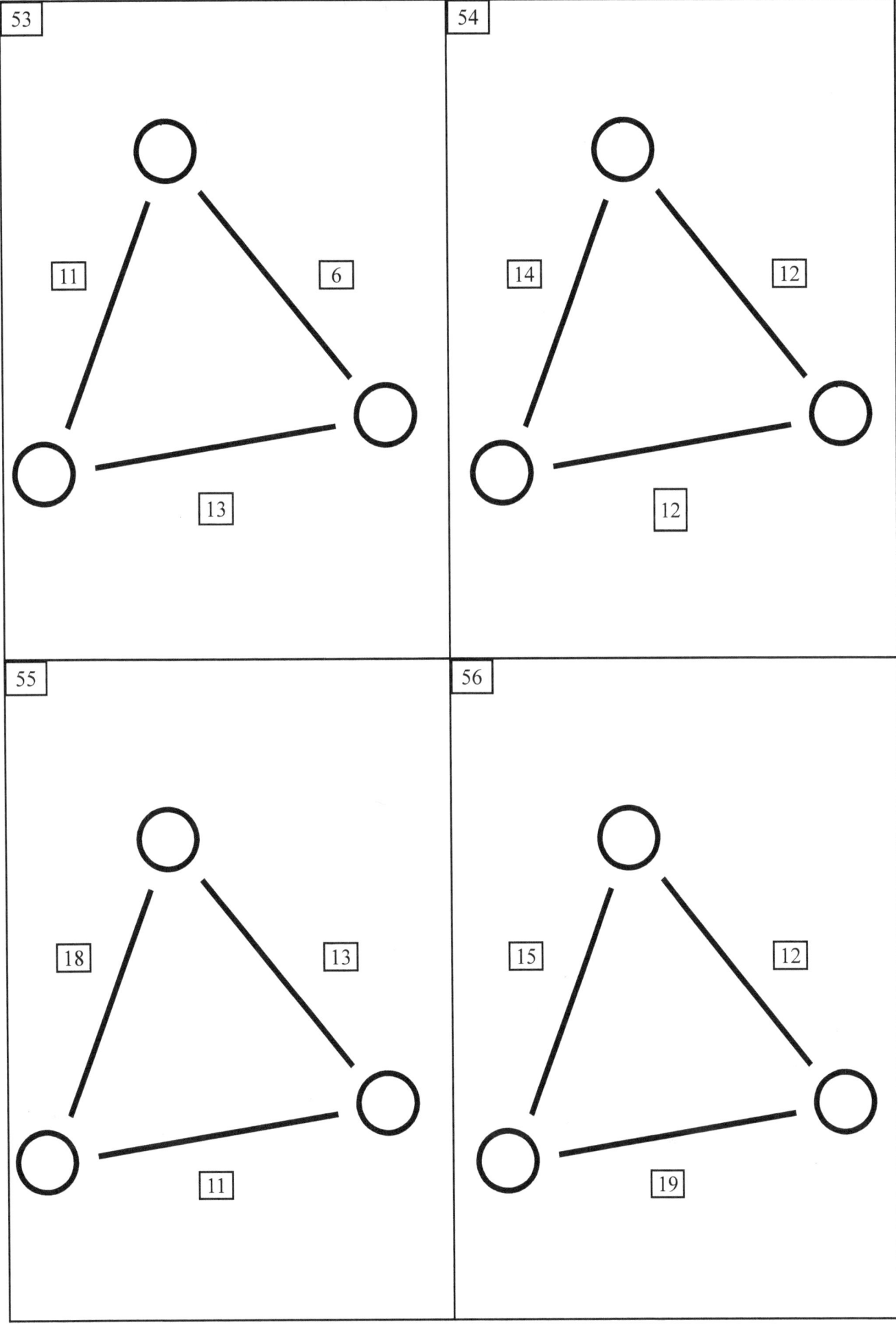

57

19    15

22

58

14    20

14

59

18    21

13

60

22    18

10

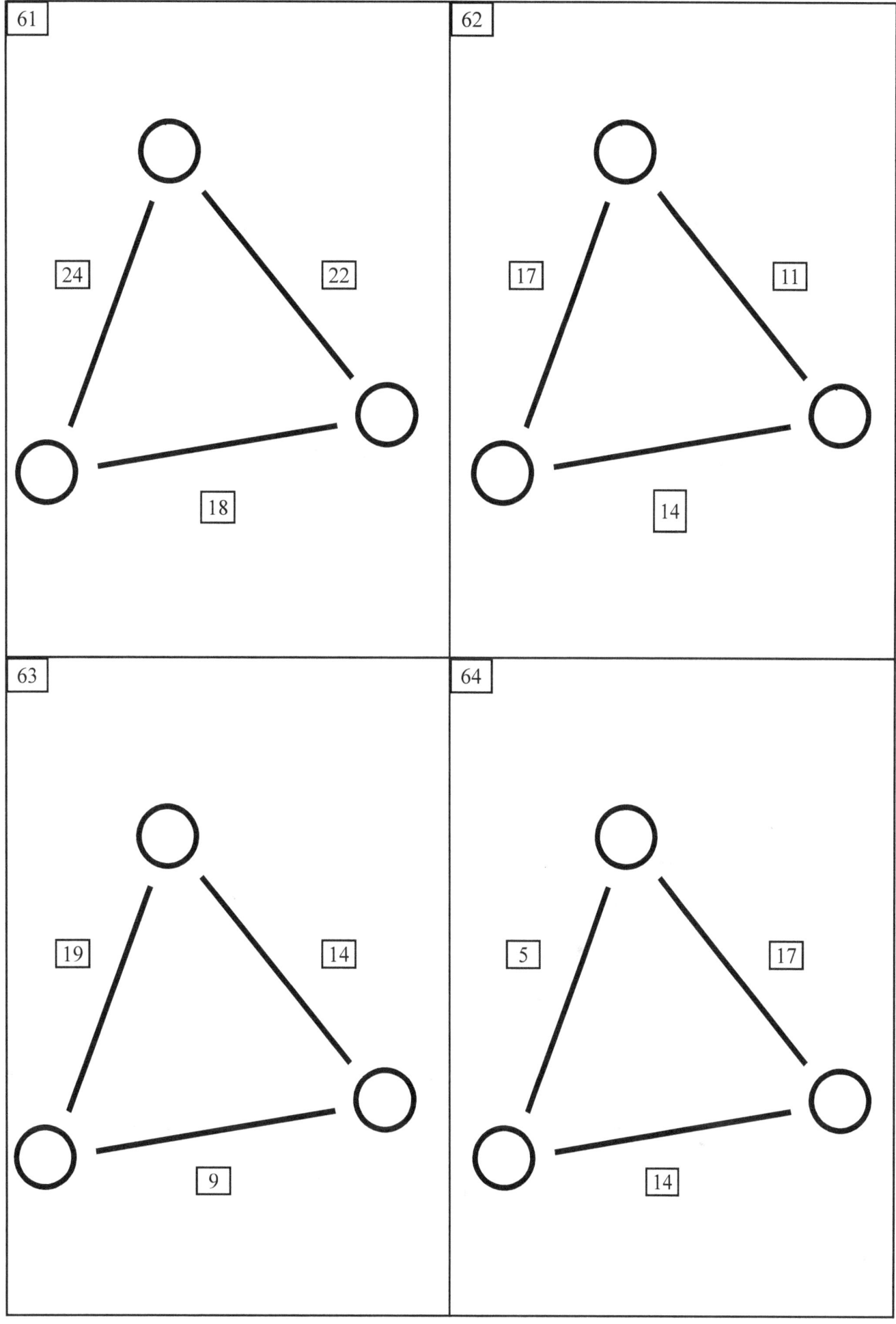

**juggle 3**

65

20  19

25

66

15  24

11

67

15  20

17

68

11  17

22

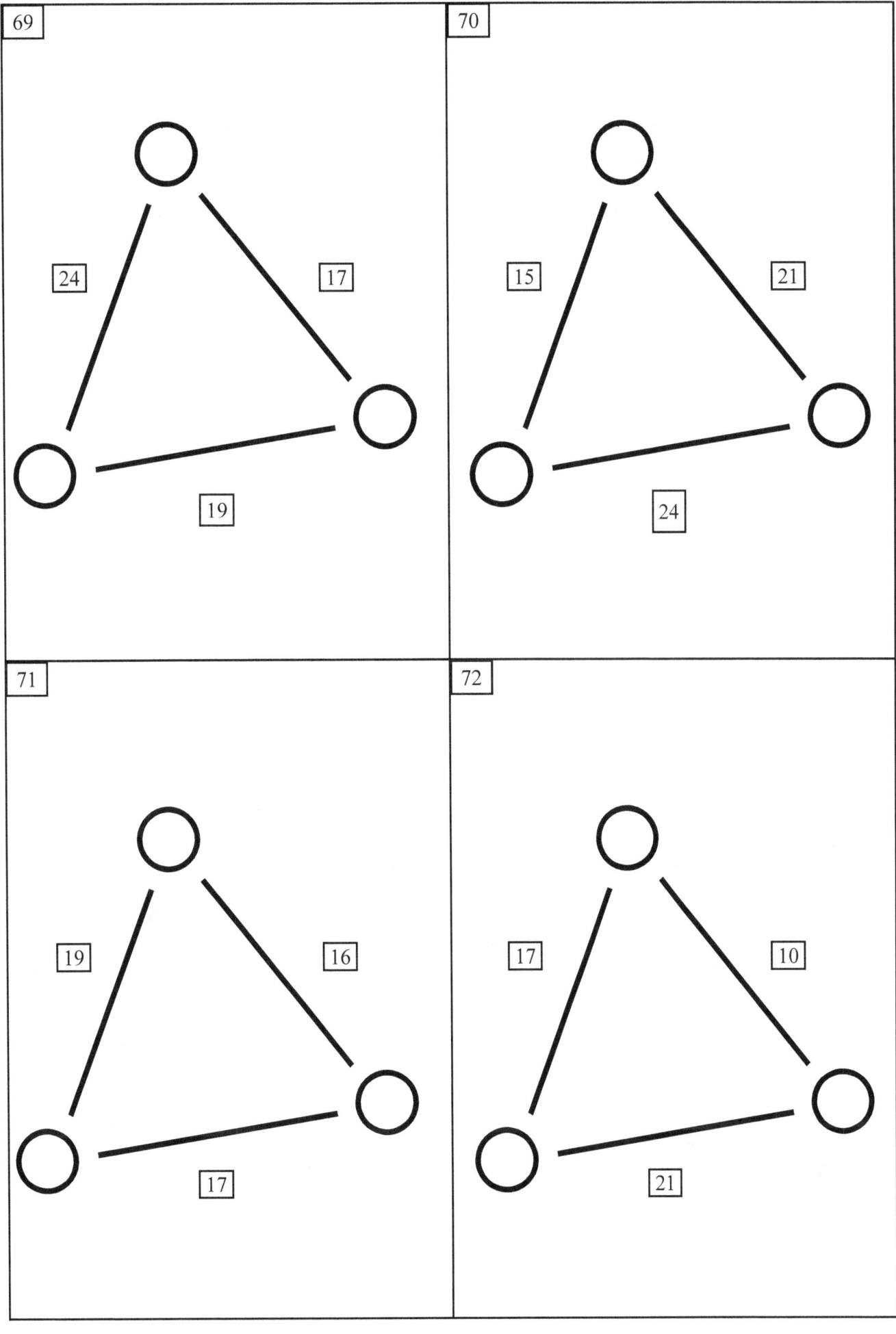

**juggle 3**

73

18 23

15

74

26 18

30

75

29 21

16

76

29 27

22

77

20  10

18

78

20  23

15

79

14  18

20

80

25  17

22

81

20    10

26

82

19    26

19

83

22    25

15

84

21    12

27

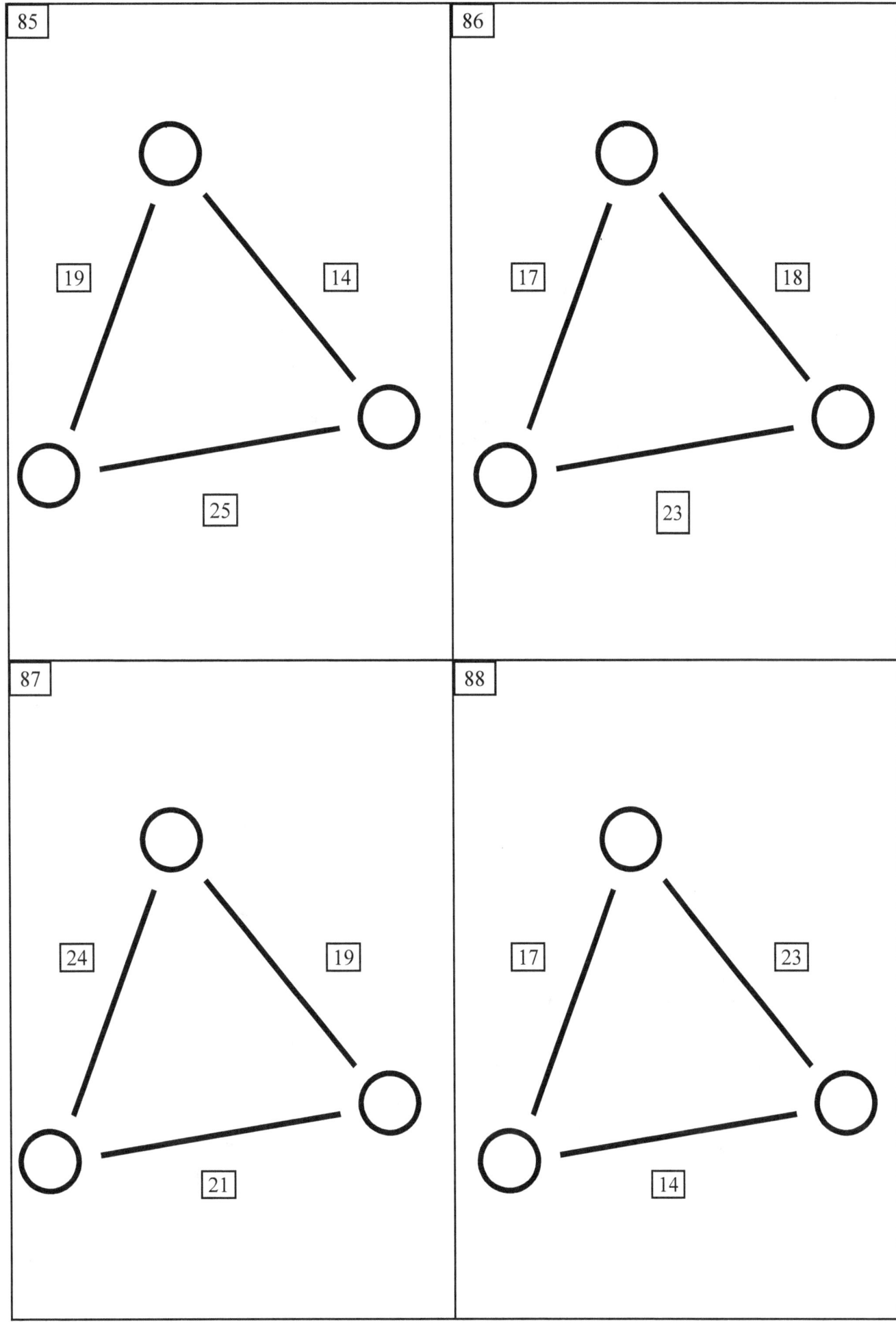

89

21 · 11 · 24

90

18 · 26 · 22

91

13 · 22 · 19

92

24 · 22 · 10

93

19    25

22

94

15    21

24

95

26    19

21

96

17    21

26

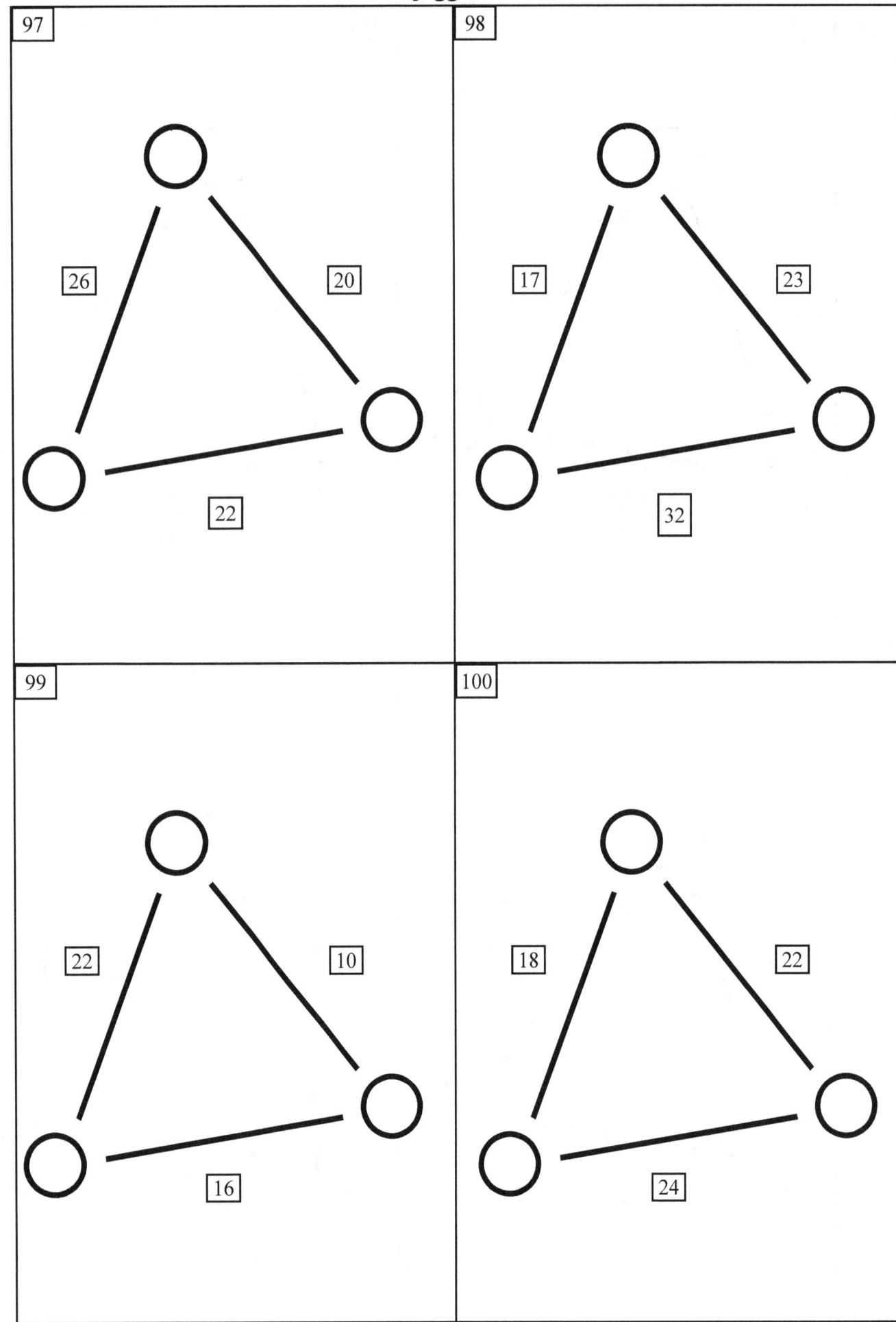

# juggle 3

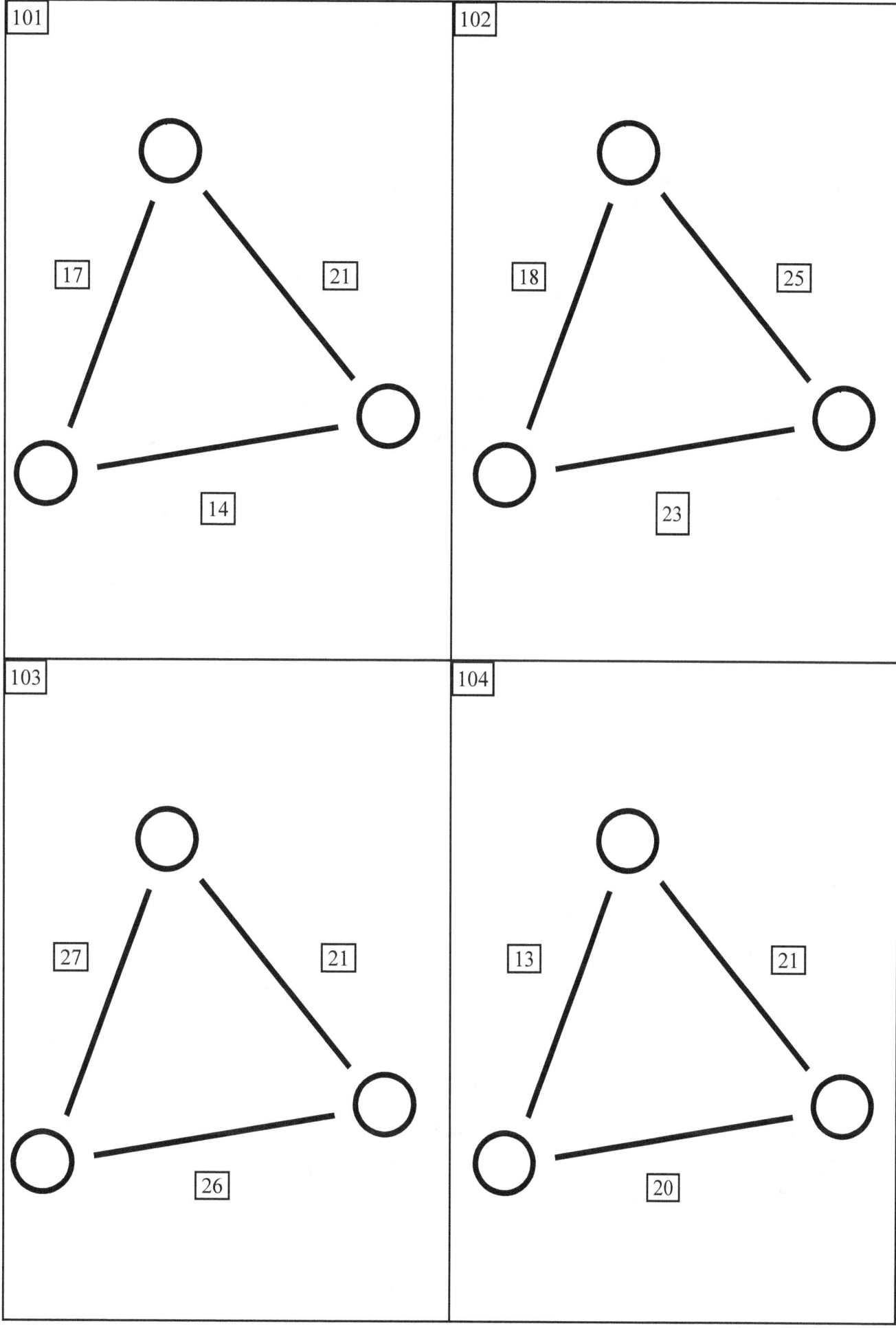

105

15  26

21

106

18  23

29

107

16  17

11

108

20  14

26

**juggle 3**

109

21   22

15

110

21   28

25

111

14   21

19

112

11   25

20

113

19    22

11

114

22    15

23

115

20    11

25

116

21    19

30

**juggle 3**

117

17    27

12

118

20    22

10

119

19    22

27

120

16    30

22

121

15    26

31

122

15    27

30

123

29    26

11

124

21    14

25

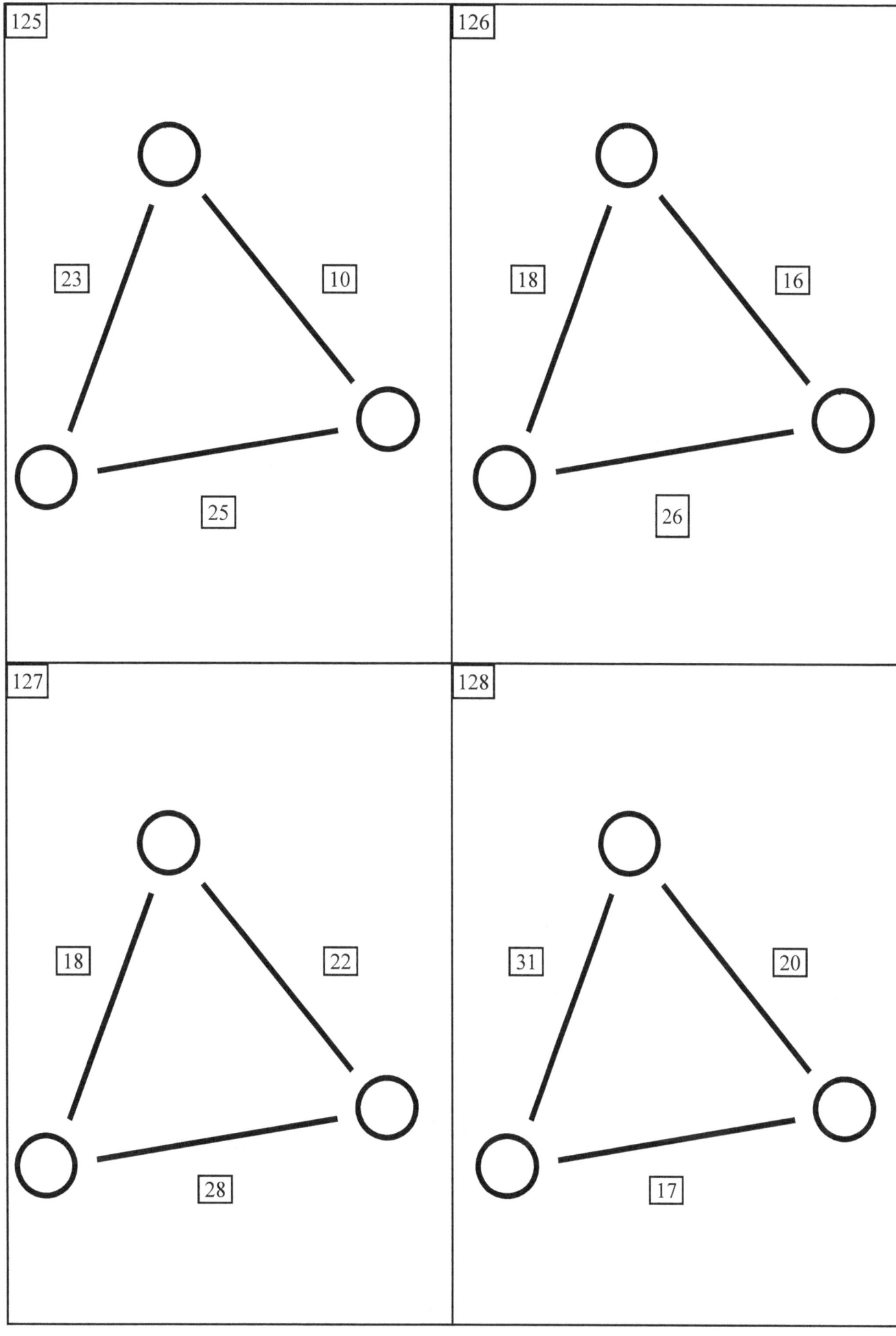

129

21    16

11

130

28    17

19

131

13    14

17

132

22    27

21

**juggle 3**

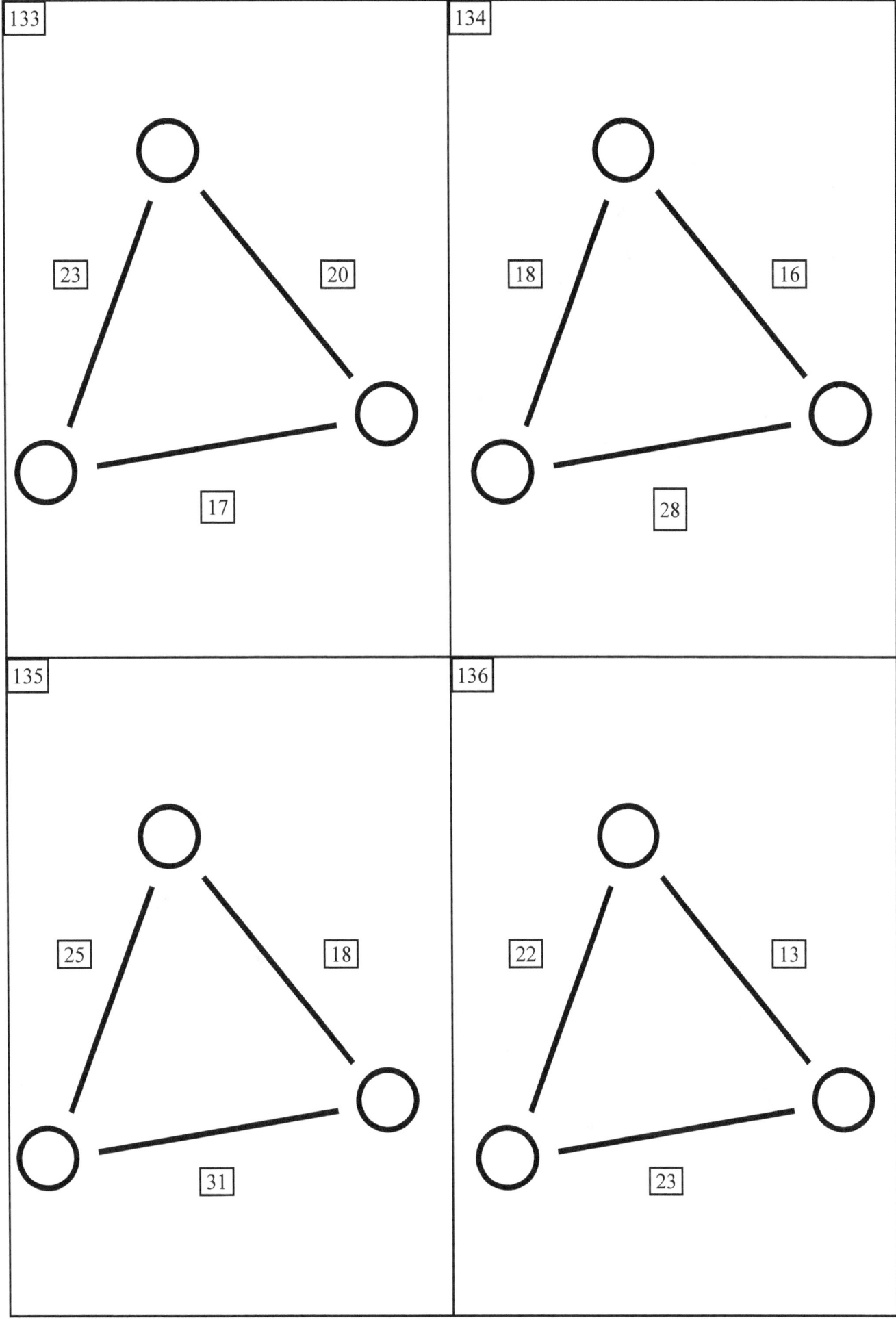

137

28    8

22

138

25    9

22

139

10    29

23

140

20    14

18

141

26      17

25

142

28      27

17

143

20      13

27

144

31      27

20

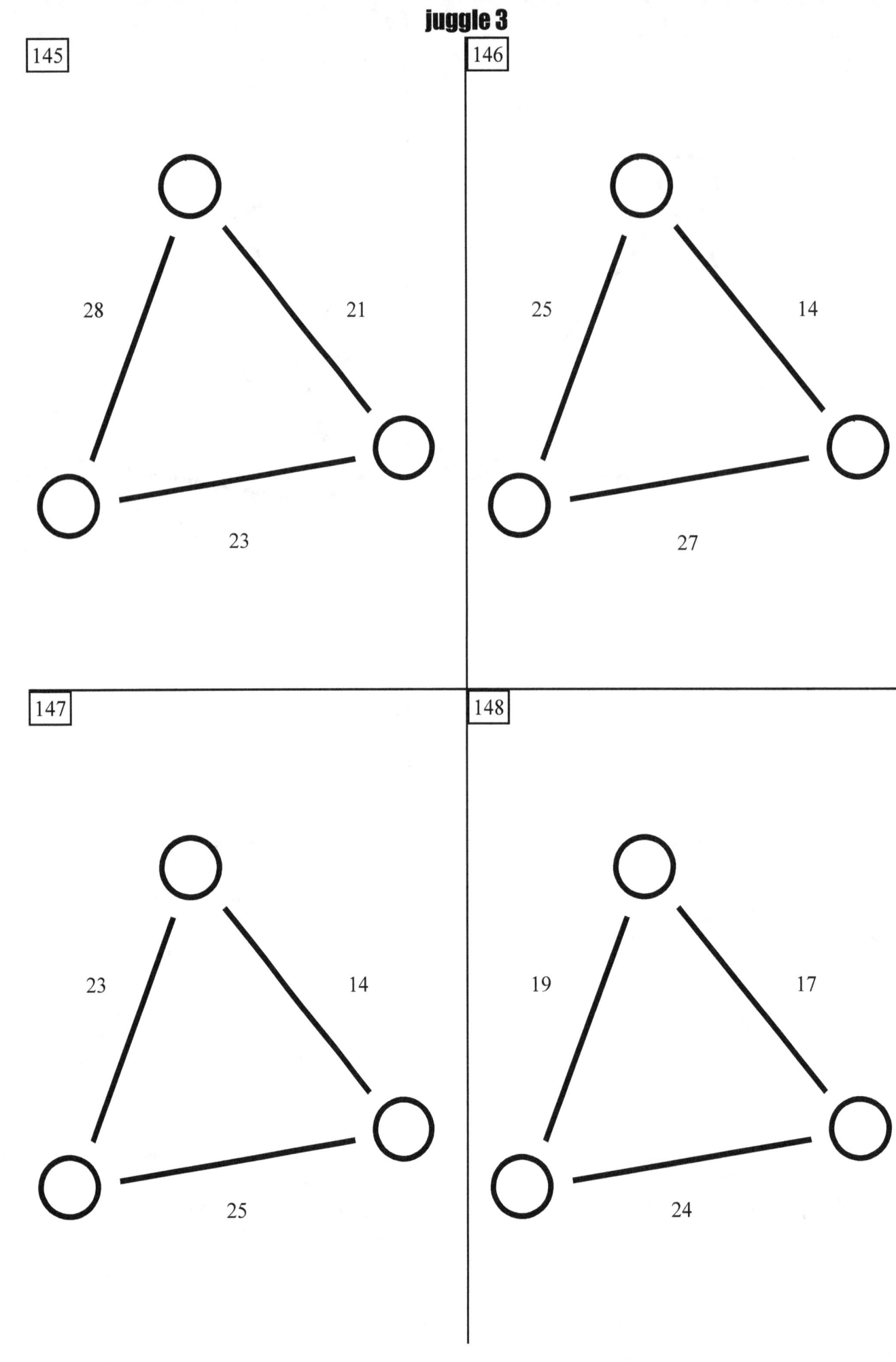

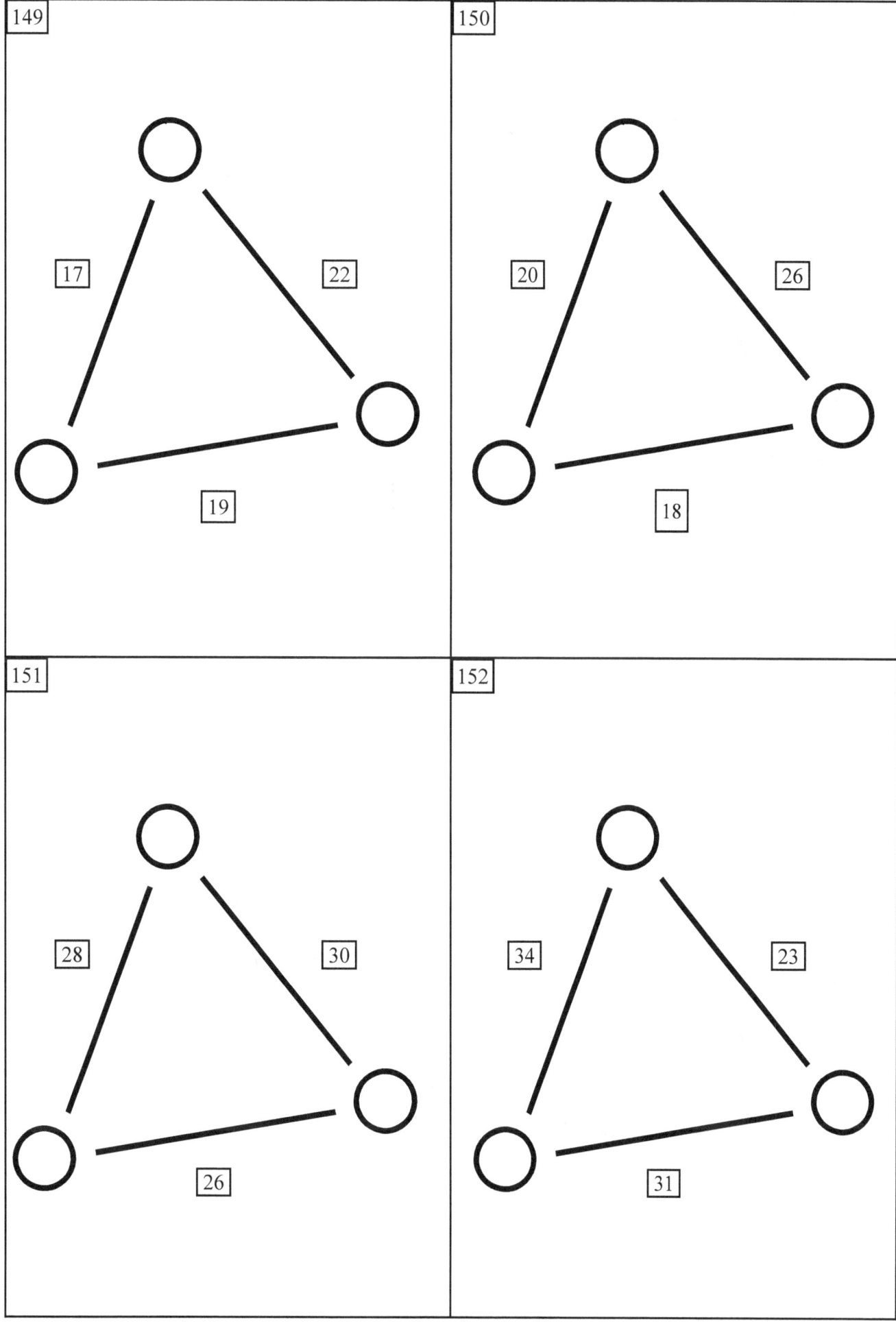

**juggle 3**

153

22
27
29

154

22
20
26

155

29
15
30

156

38
24
28

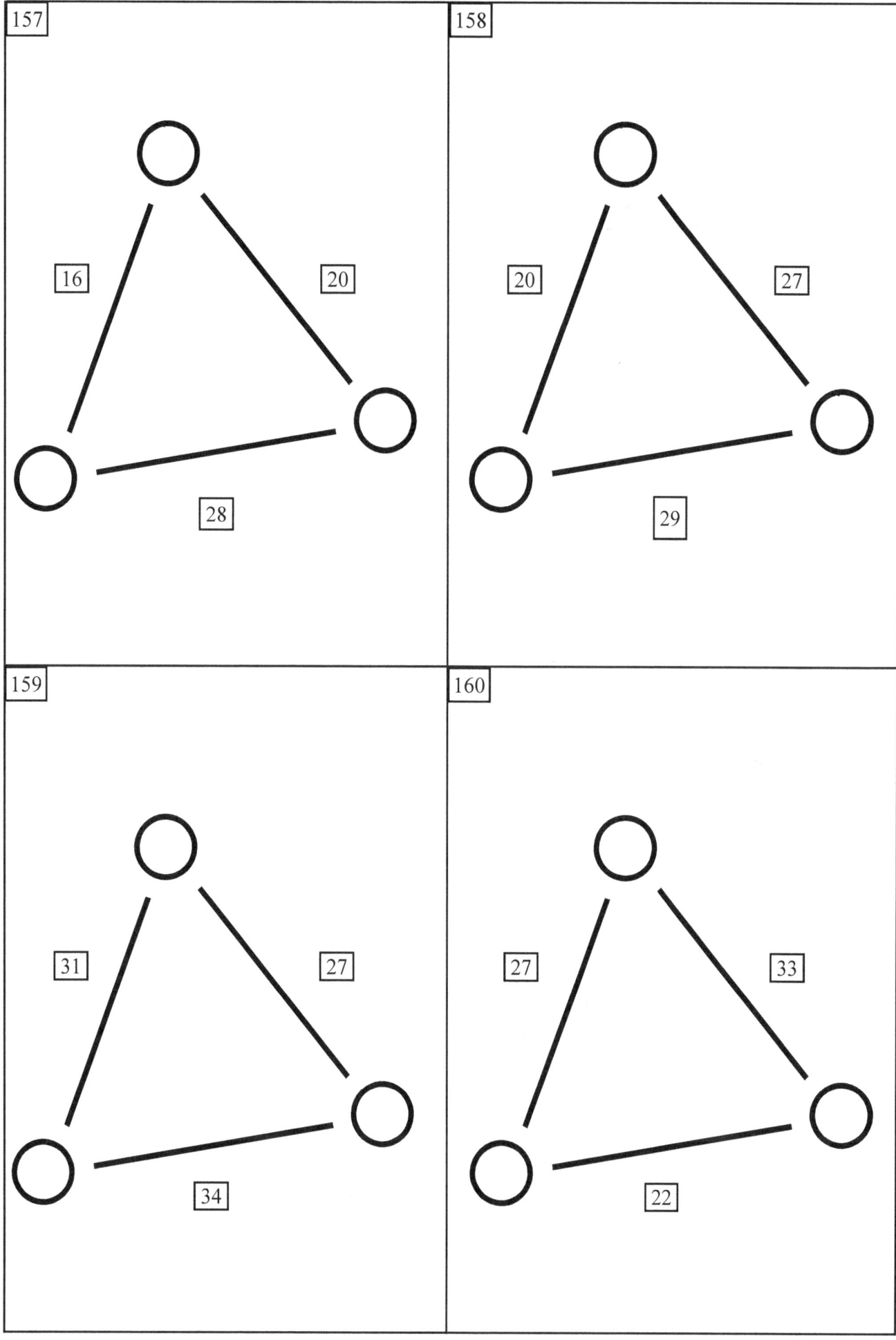

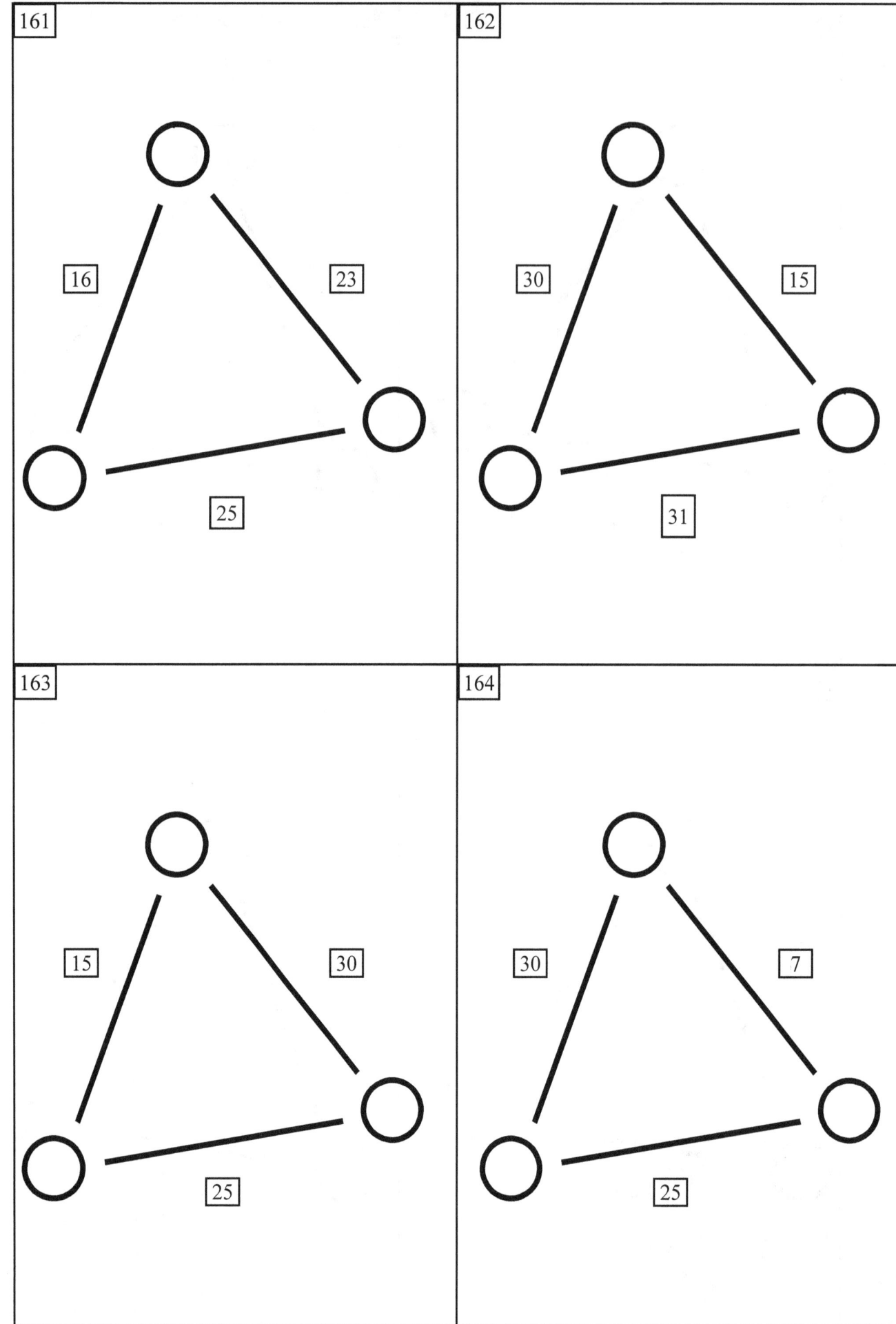

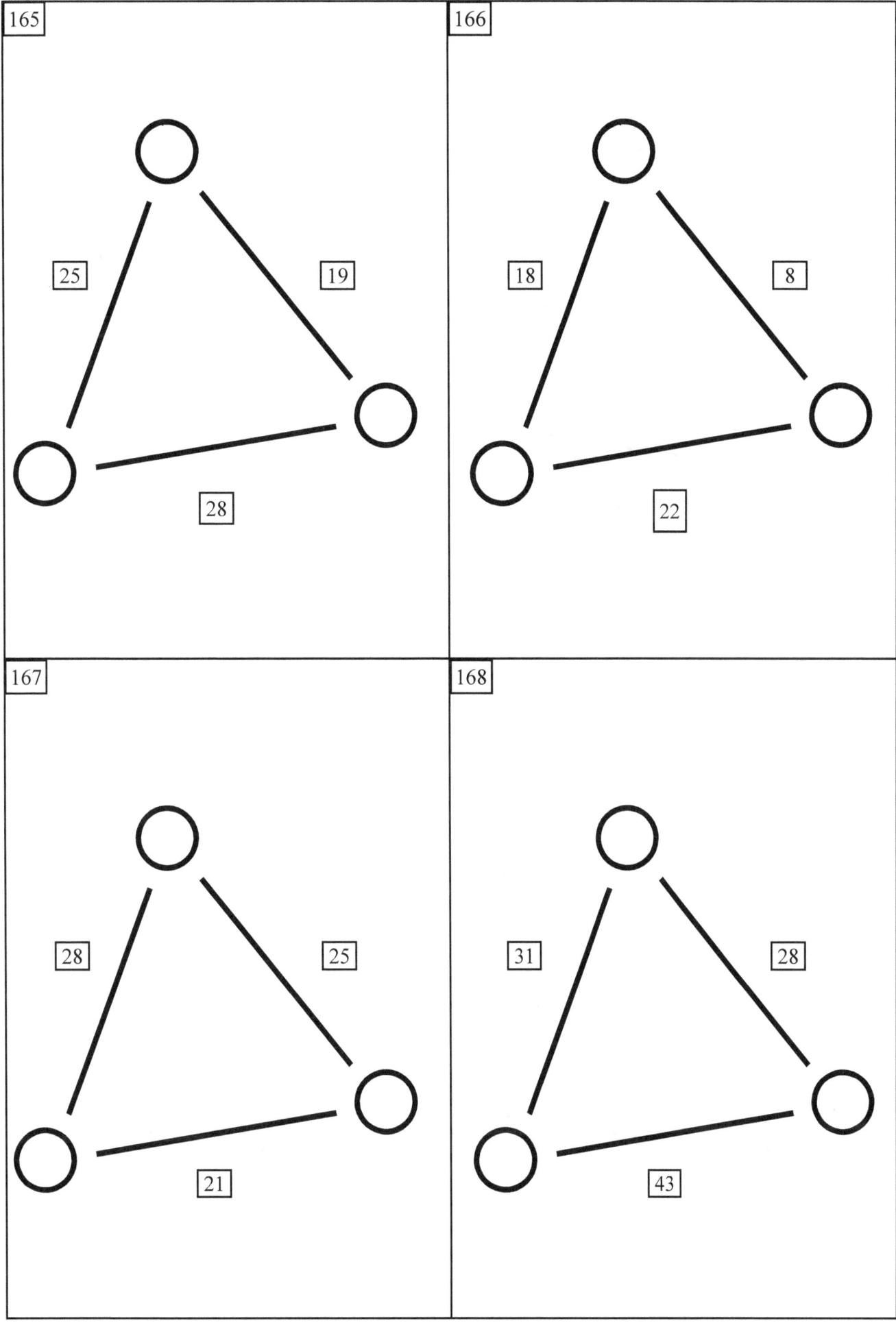

169

22    11

9

170

21    33

14

171

31    25

32

172

26    29

21

**juggle 3**

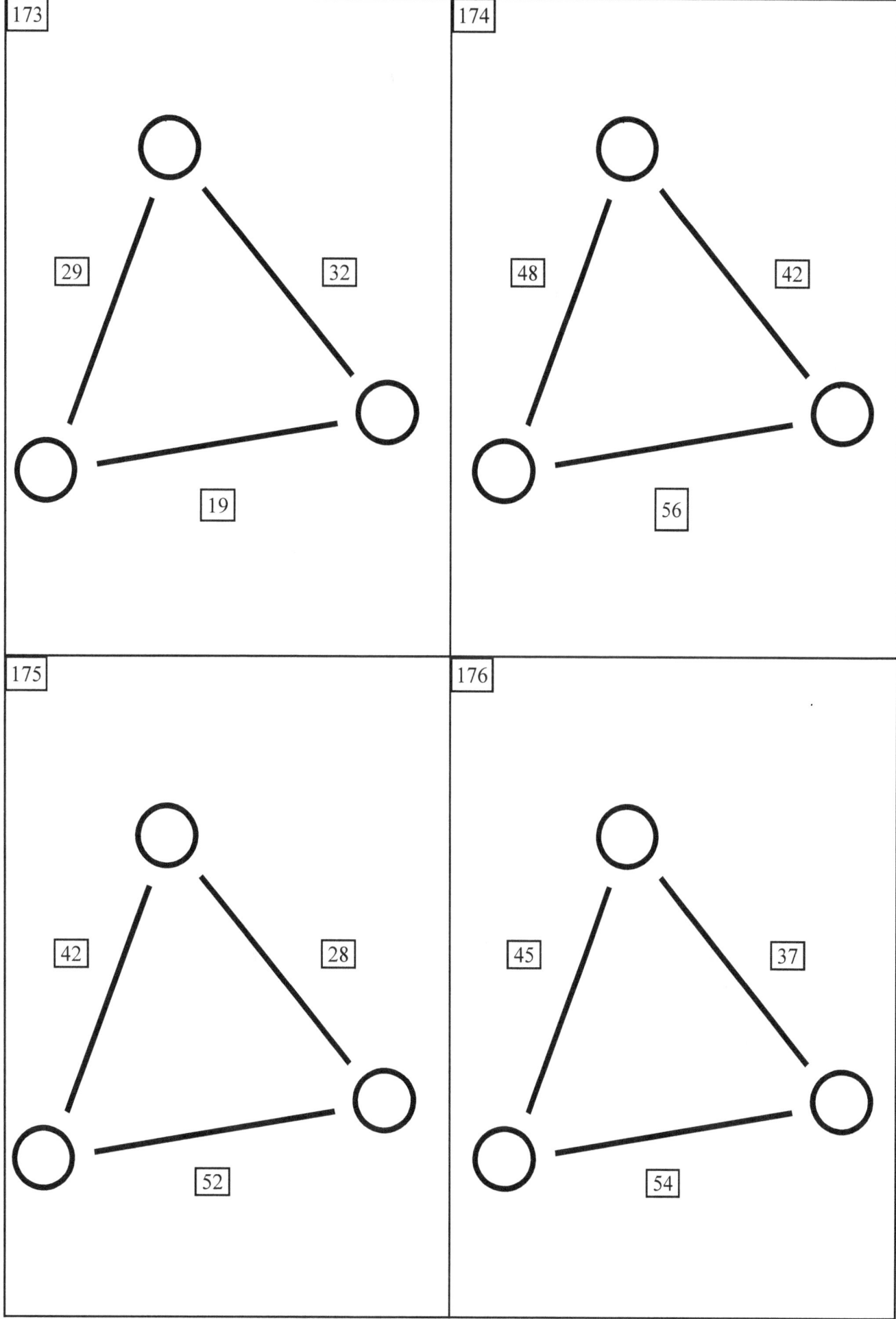

**juggle 3**

177

61
27
50

178

43
51
34

179

59
32
57

180

71
50
33

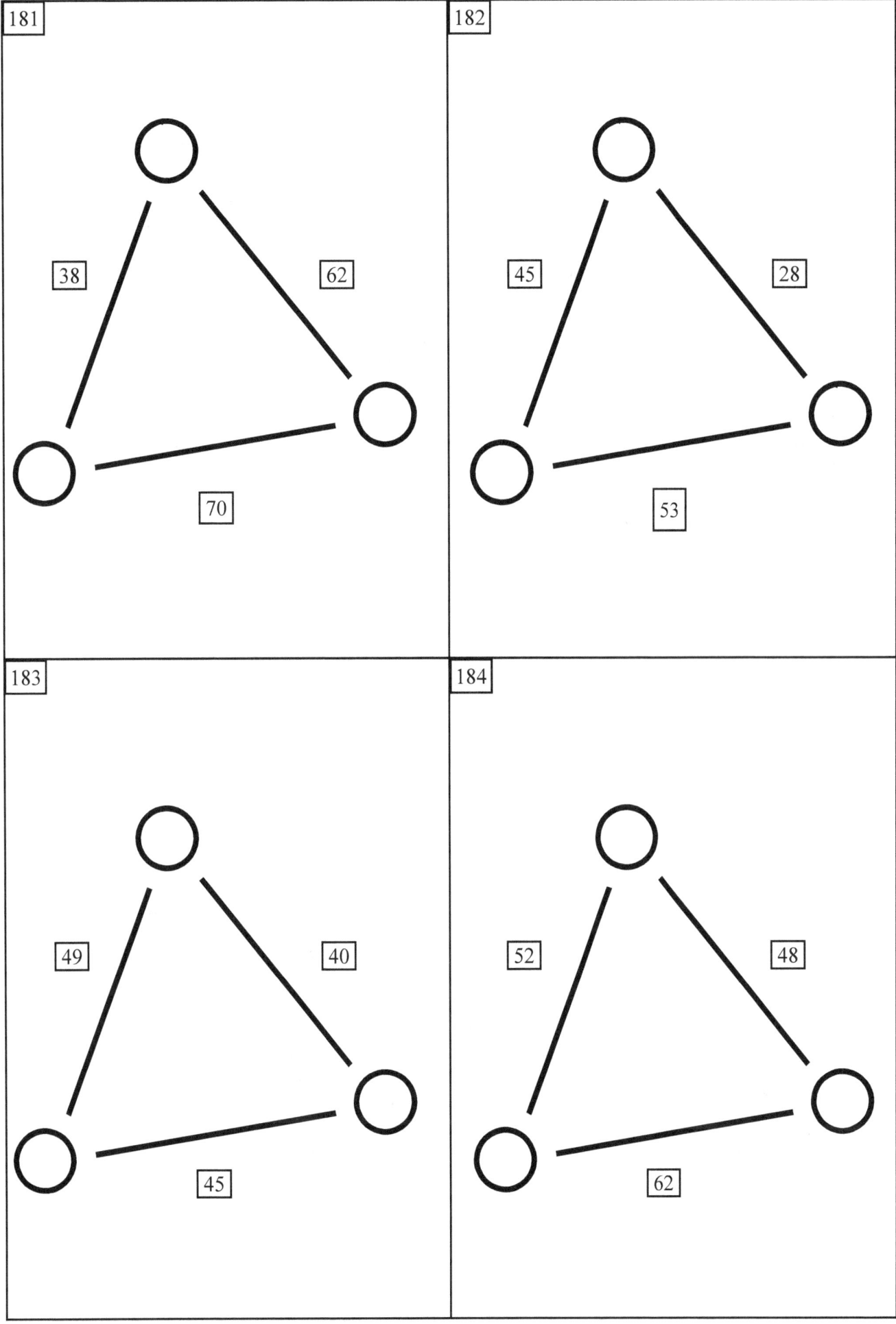

185

35    63

52

186

56    57

33

187

43    45

34

188

48    33

71

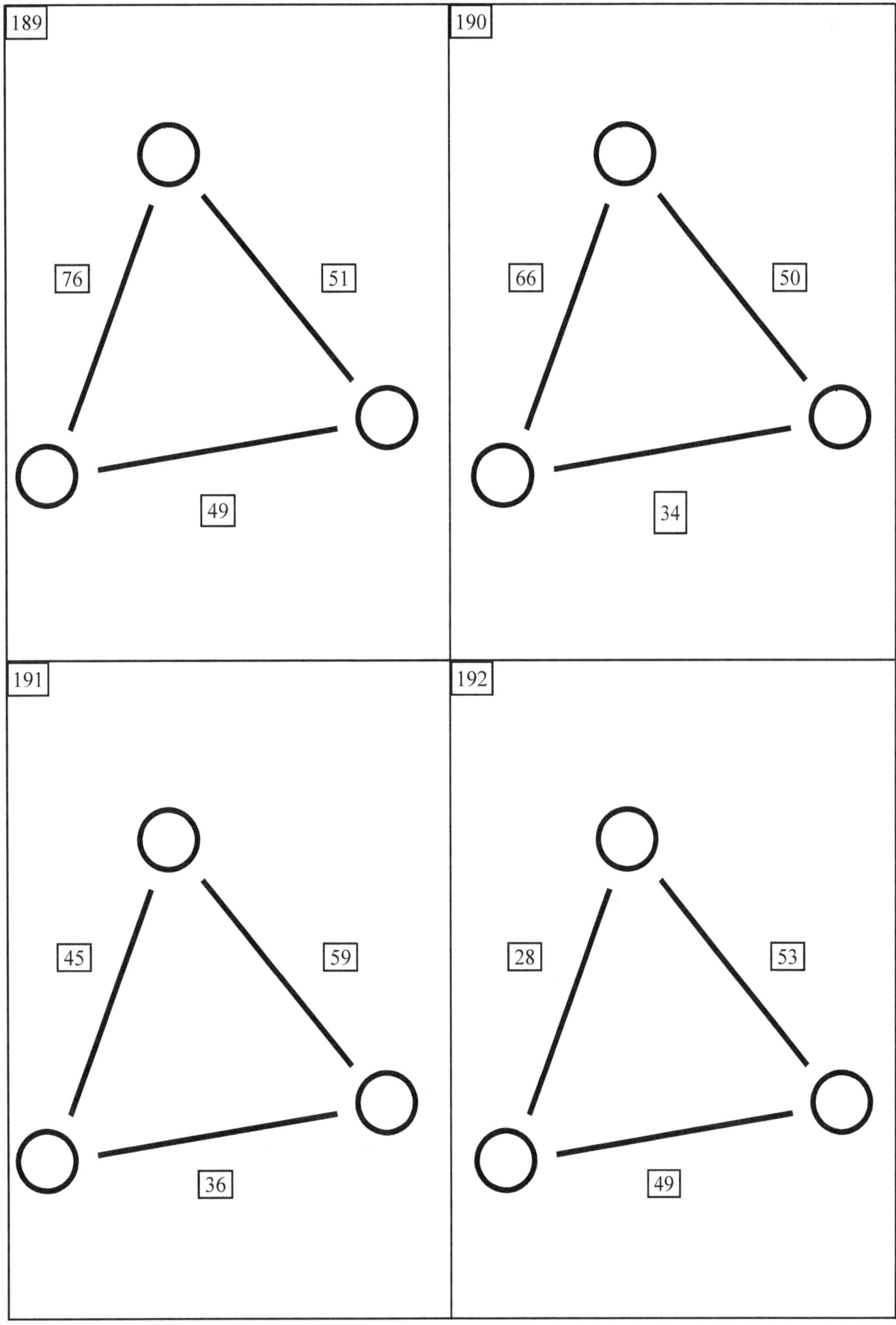

193

35   45

44

194

55   36

41

195

40   41

33

196

46   47

27

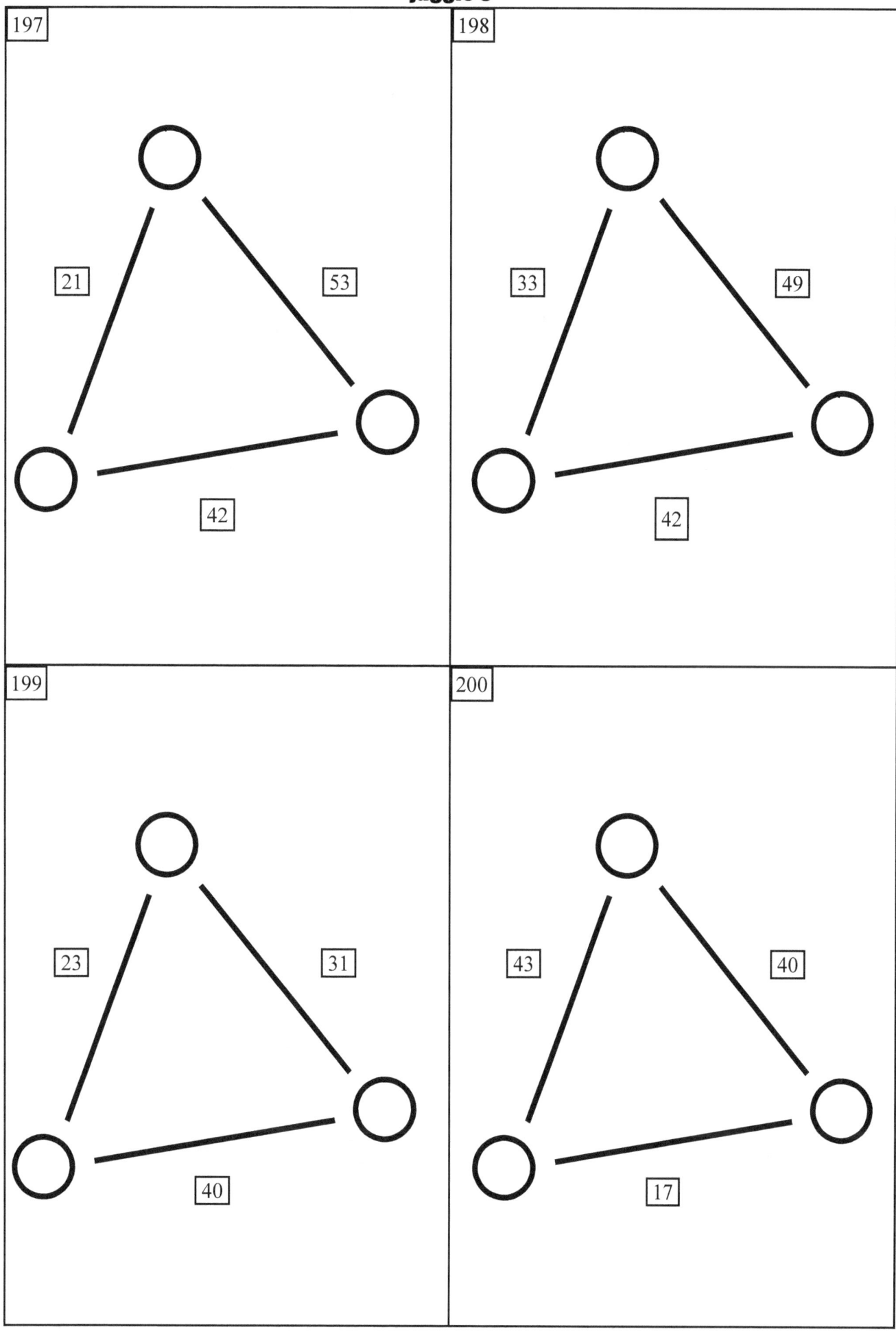

**juggle 3**

201

48   55

17

202

50   72

36

203

29   71

66

204

61   53

82

**juggle 3**

205

42
27
51

206

15
40
39

207

34
21
49

208

41
26
45

**juggle 3**

209

41    26

43

210

44    25

35

211

54    32

60

212

99    50

63

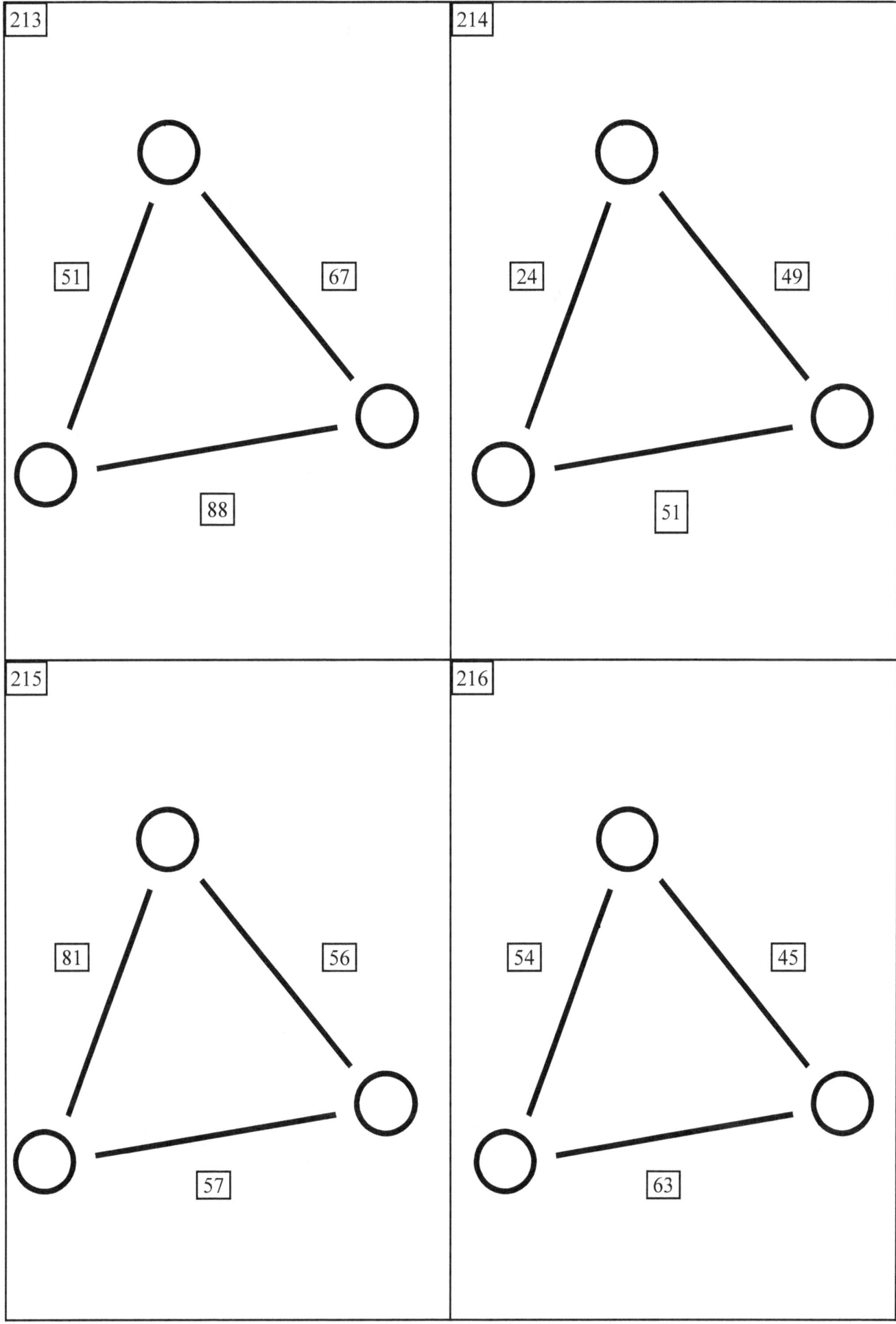

217

74   52

36

218

52   58

22

219

60   36

72

220

18   34

30

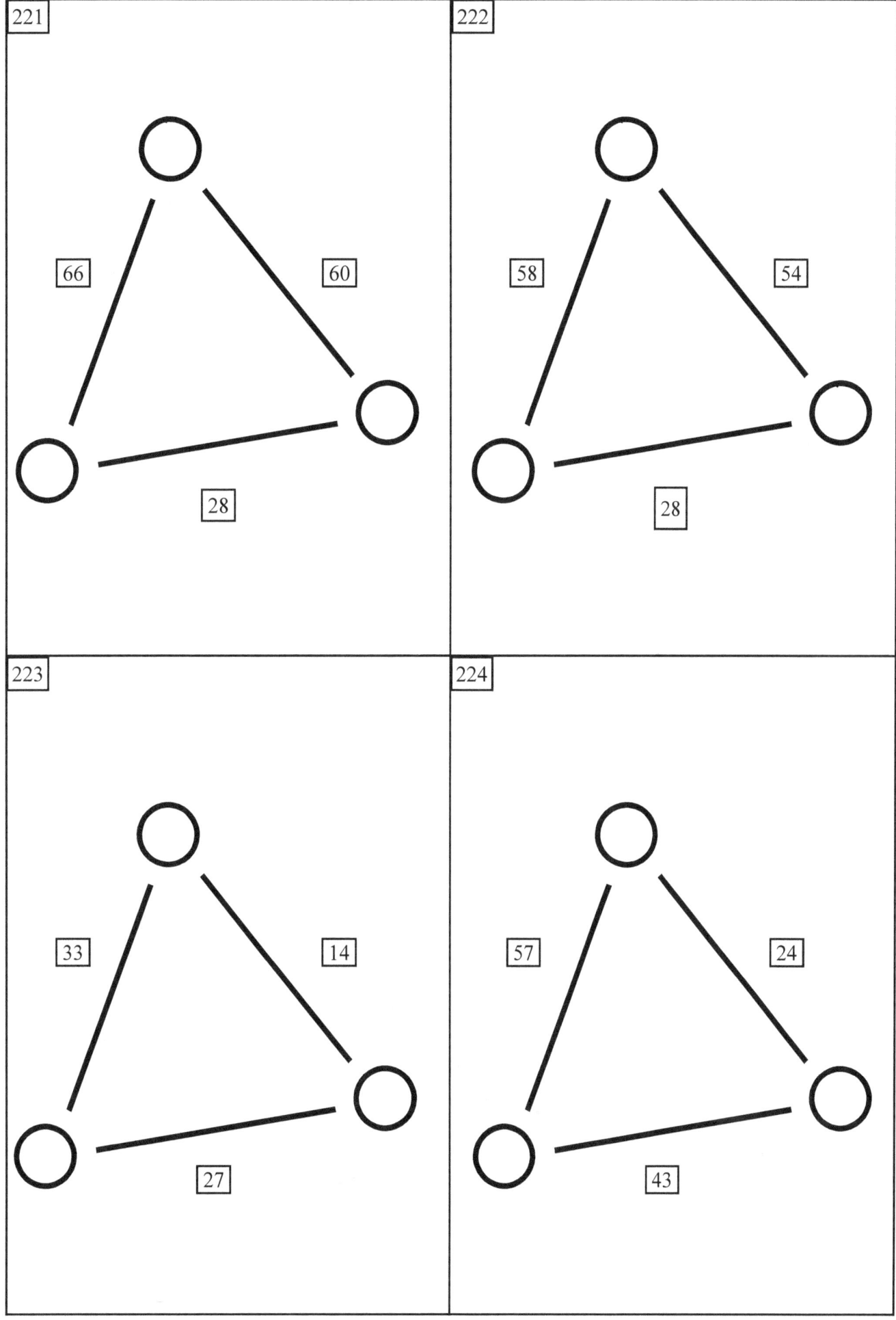

**juggle 3**

225

21  59

50

226

100  42

64

227

53  87

38

228

44  43

13

# Juggle 3 - Subtract

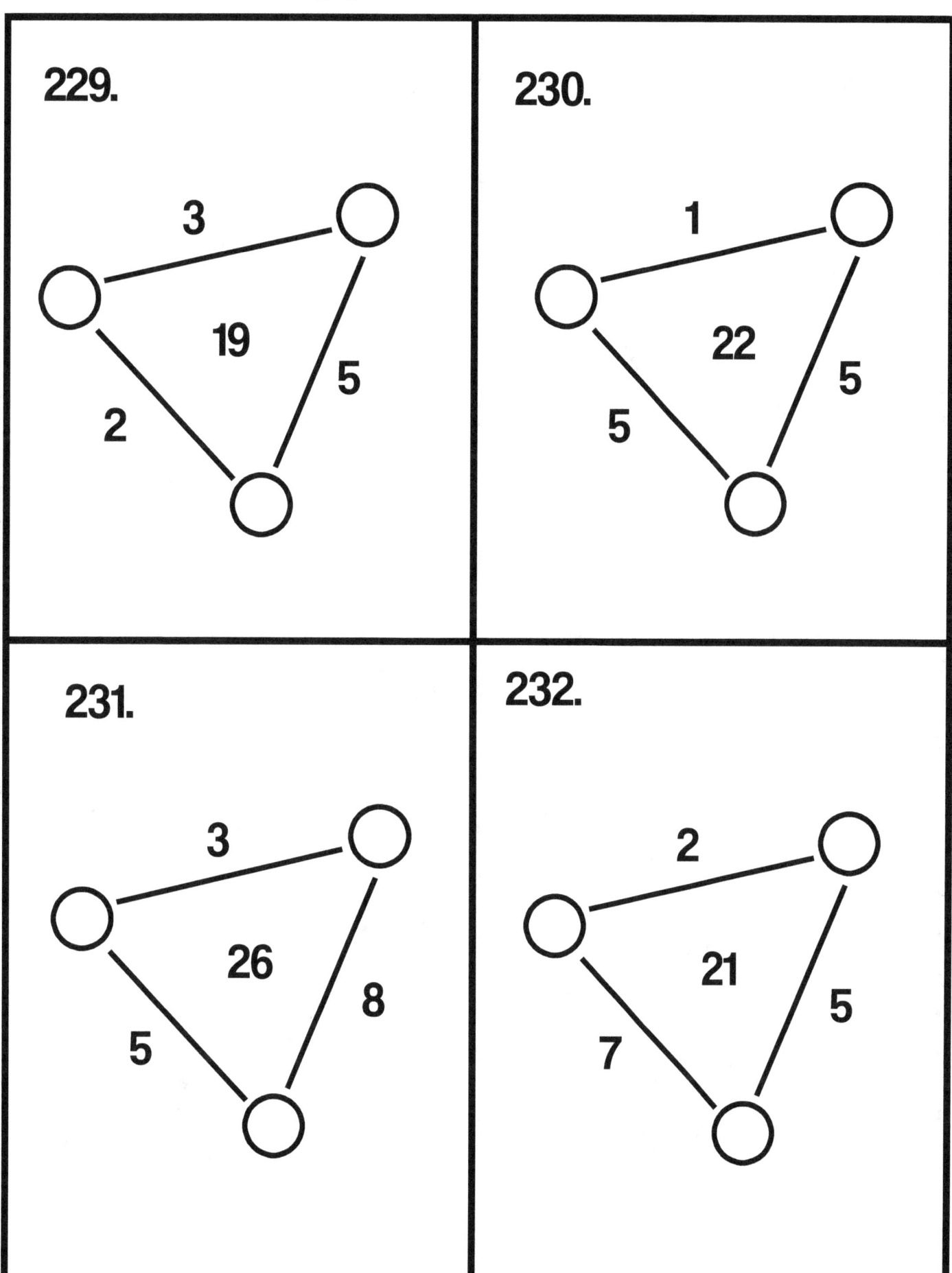

229.
3
19
5
2

230.
1
22
5
5

231.
3
26
8
5

232.
2
21
5
7

# Juggle 3 - Subtract

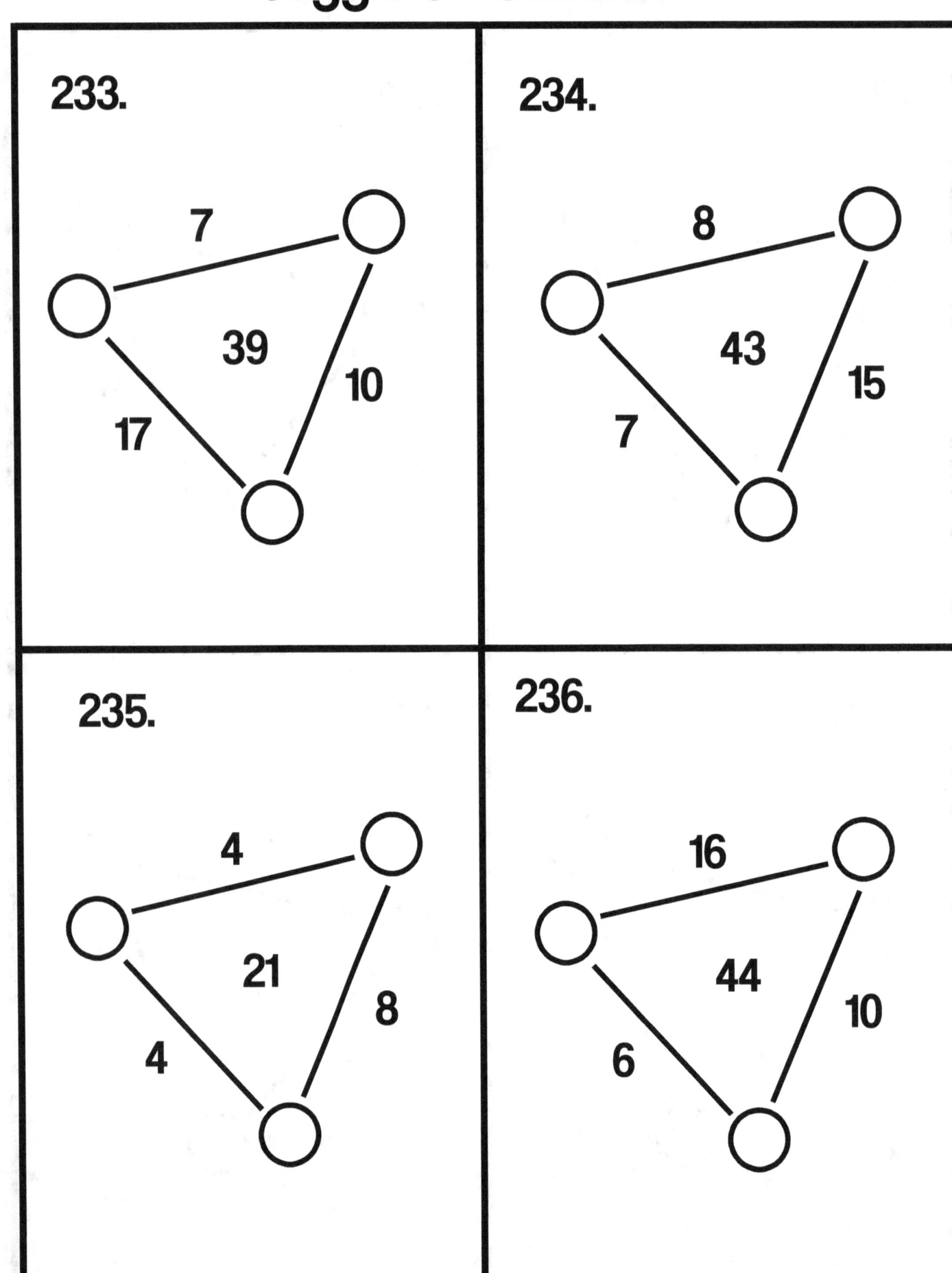

233.

7   39   10   17

234.

8   43   15   7

235.

4   21   8   4

236.

16   44   10   6

# Juggle 3 - Subtract

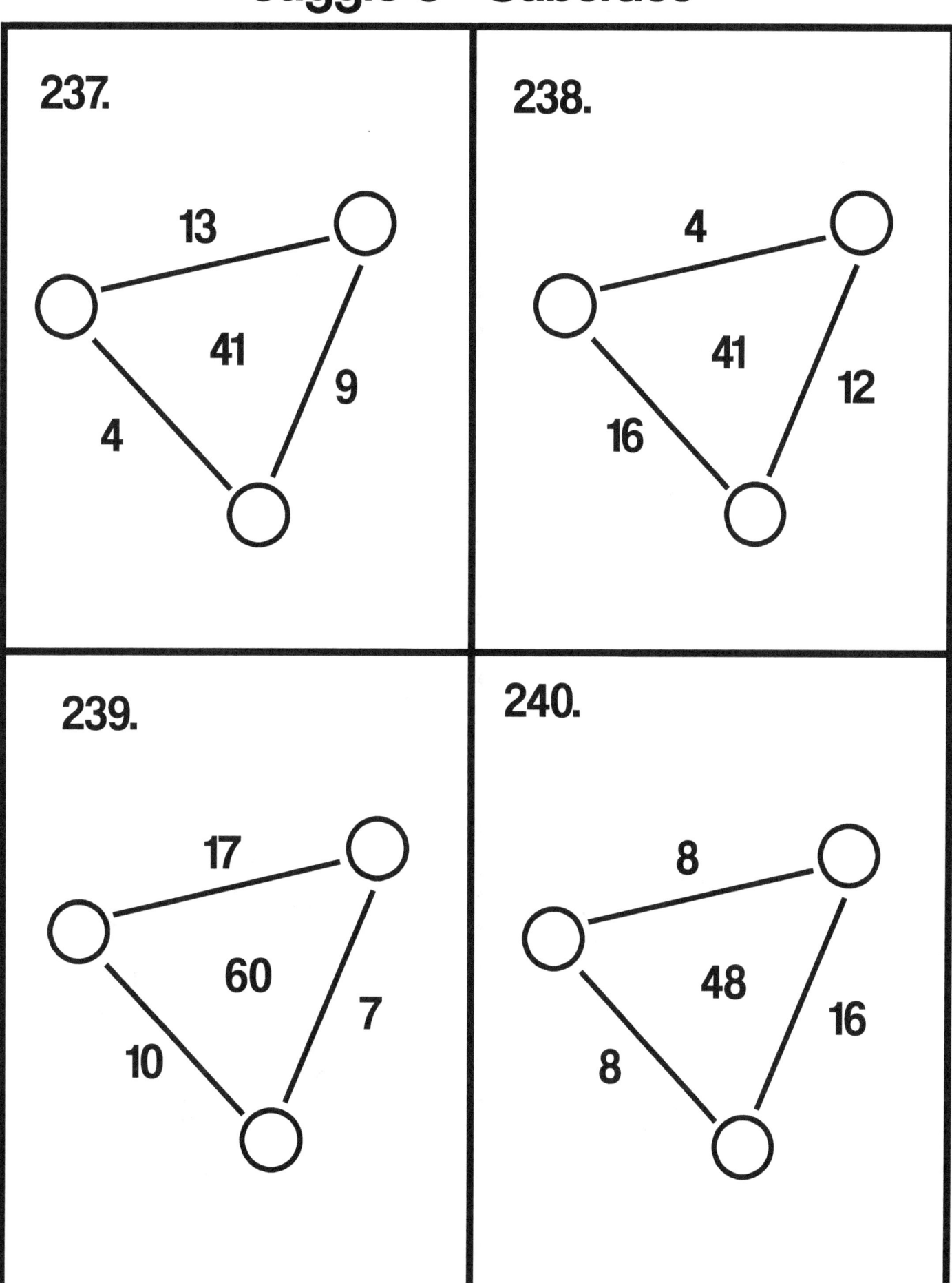

# Juggle 3 - Subtract

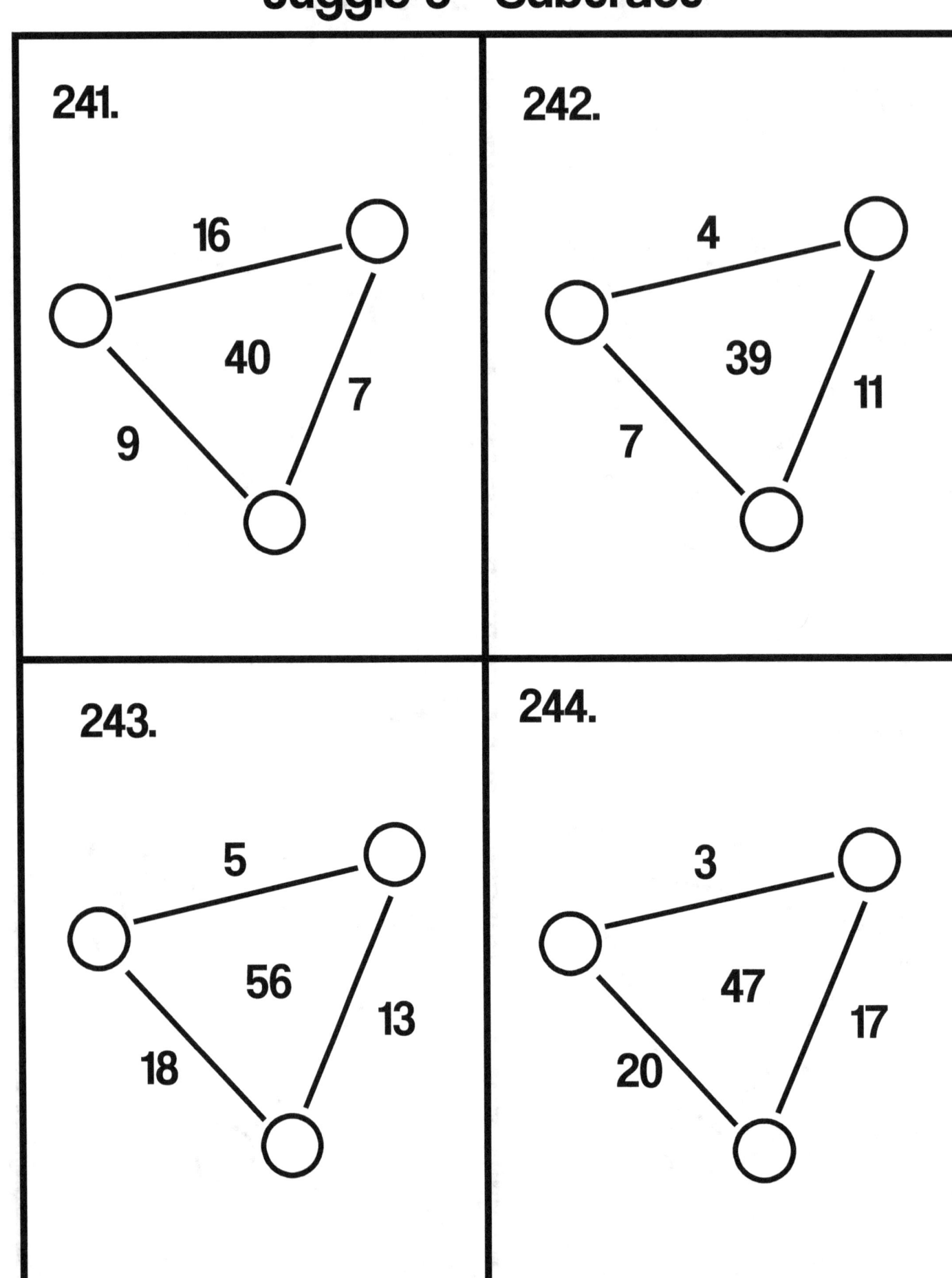

241.

16

40

7

9

242.

4

39

11

7

243.

5

56

13

18

244.

3

47

17

20

# Juggle 3 - Subtract

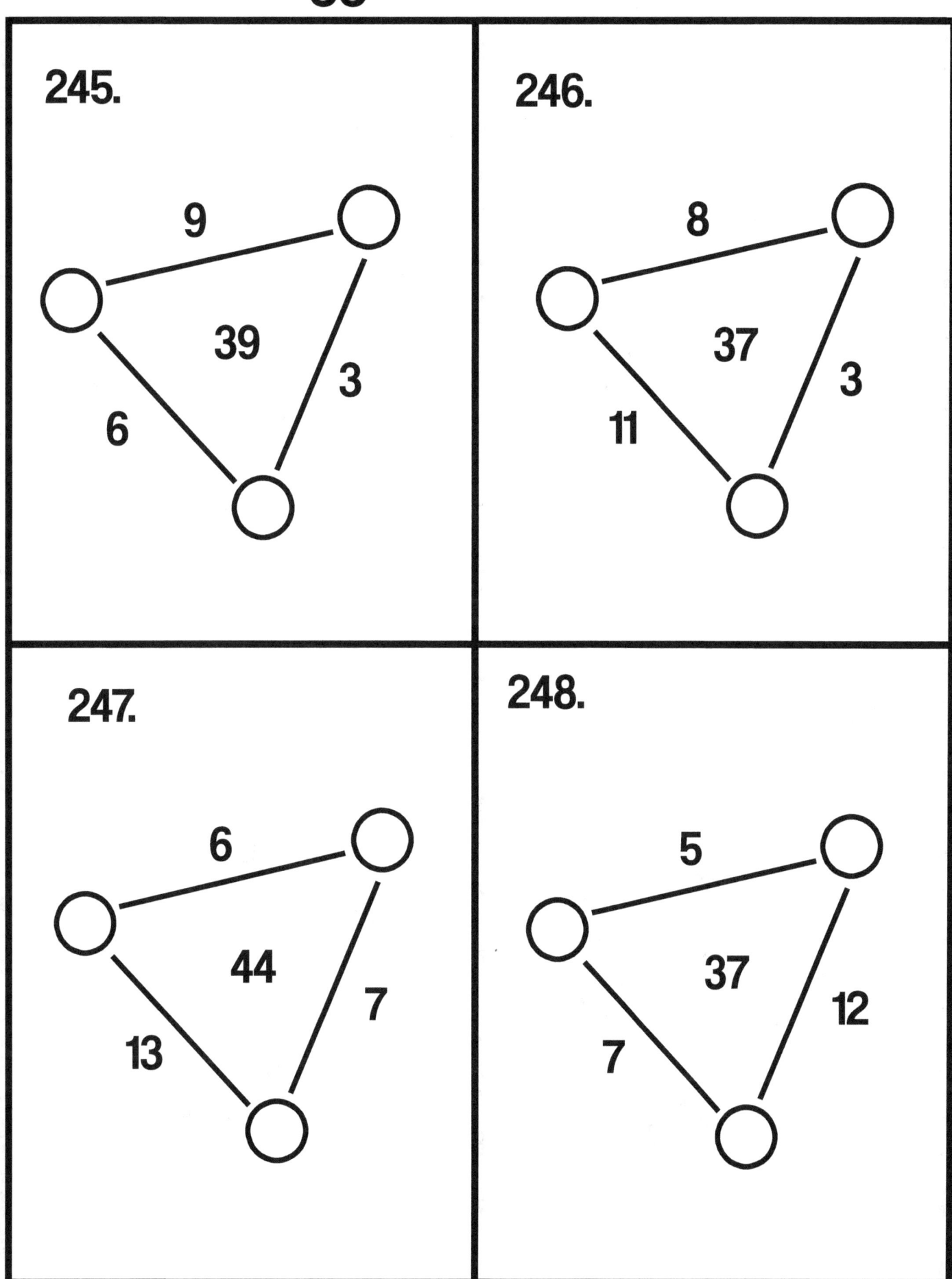

245.

9
39
3
6

246.

8
37
3
11

247.

6
44
7
13

248.

5
37
12
7

# Juggle 3 - Subtract

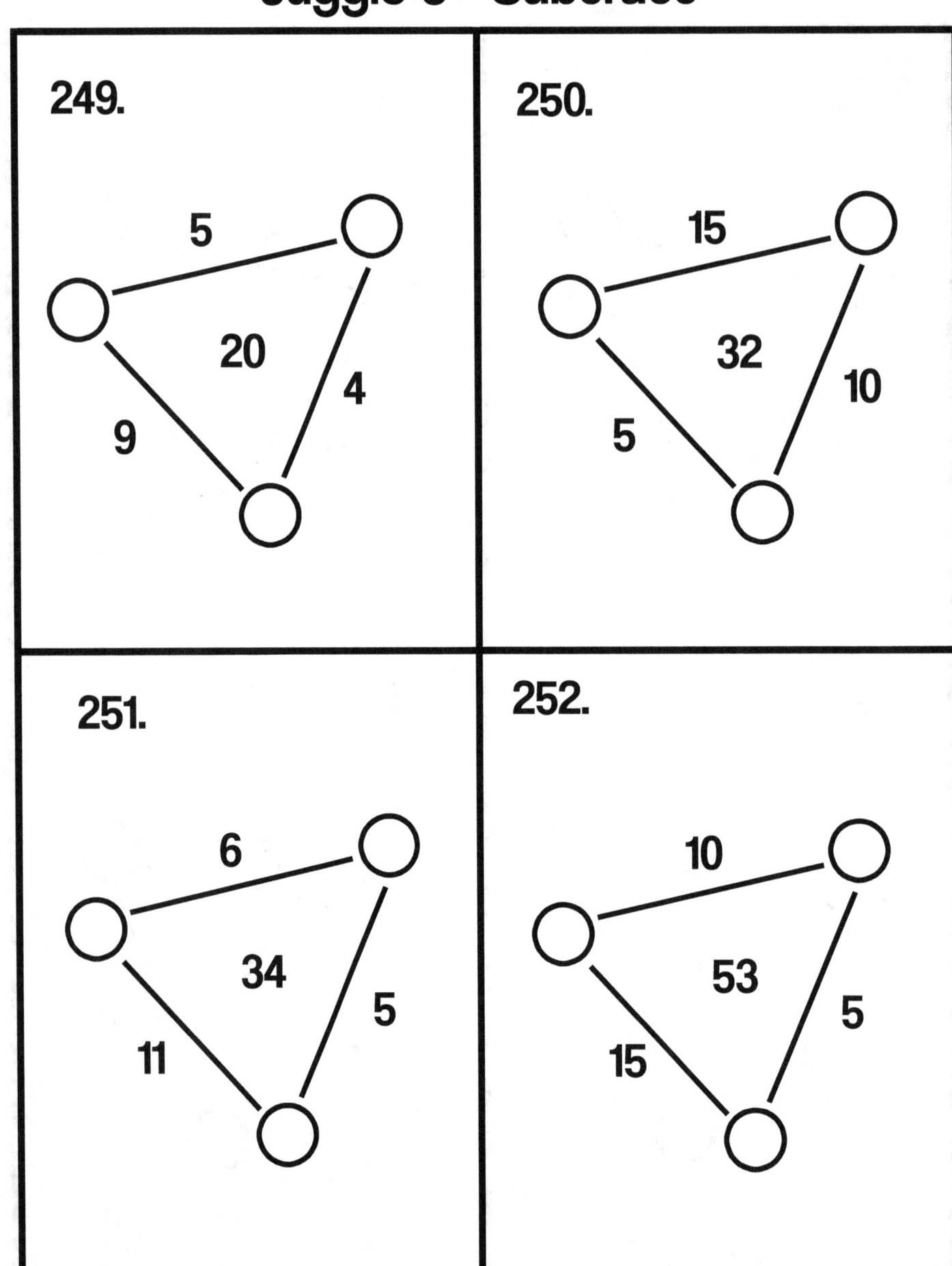

# Juggle 3 - Subtract

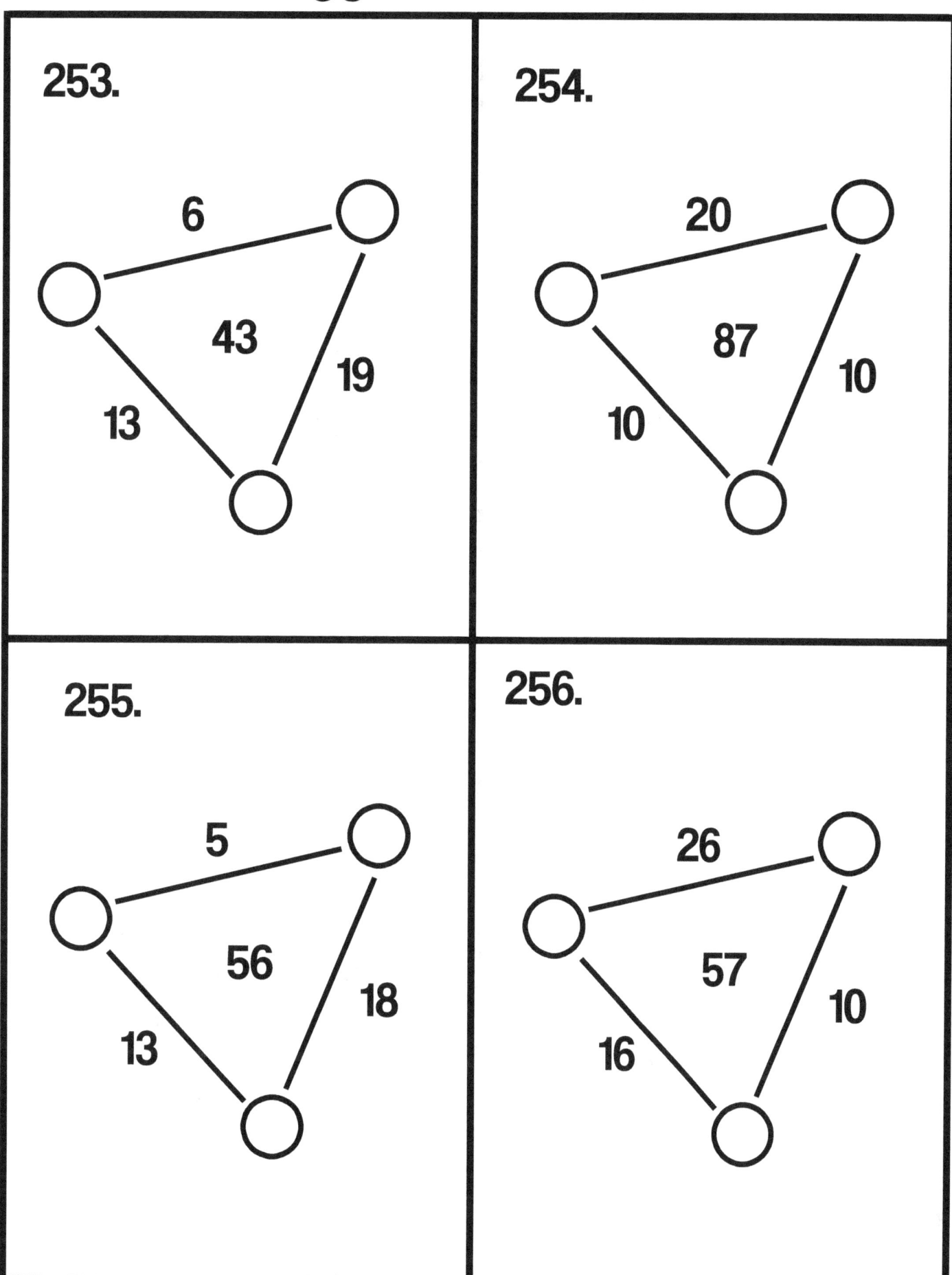

253.

6

43

19

13

254.

20

87

10

10

255.

5

56

18

13

256.

26

57

10

16

# Juggle 3 - Subtract

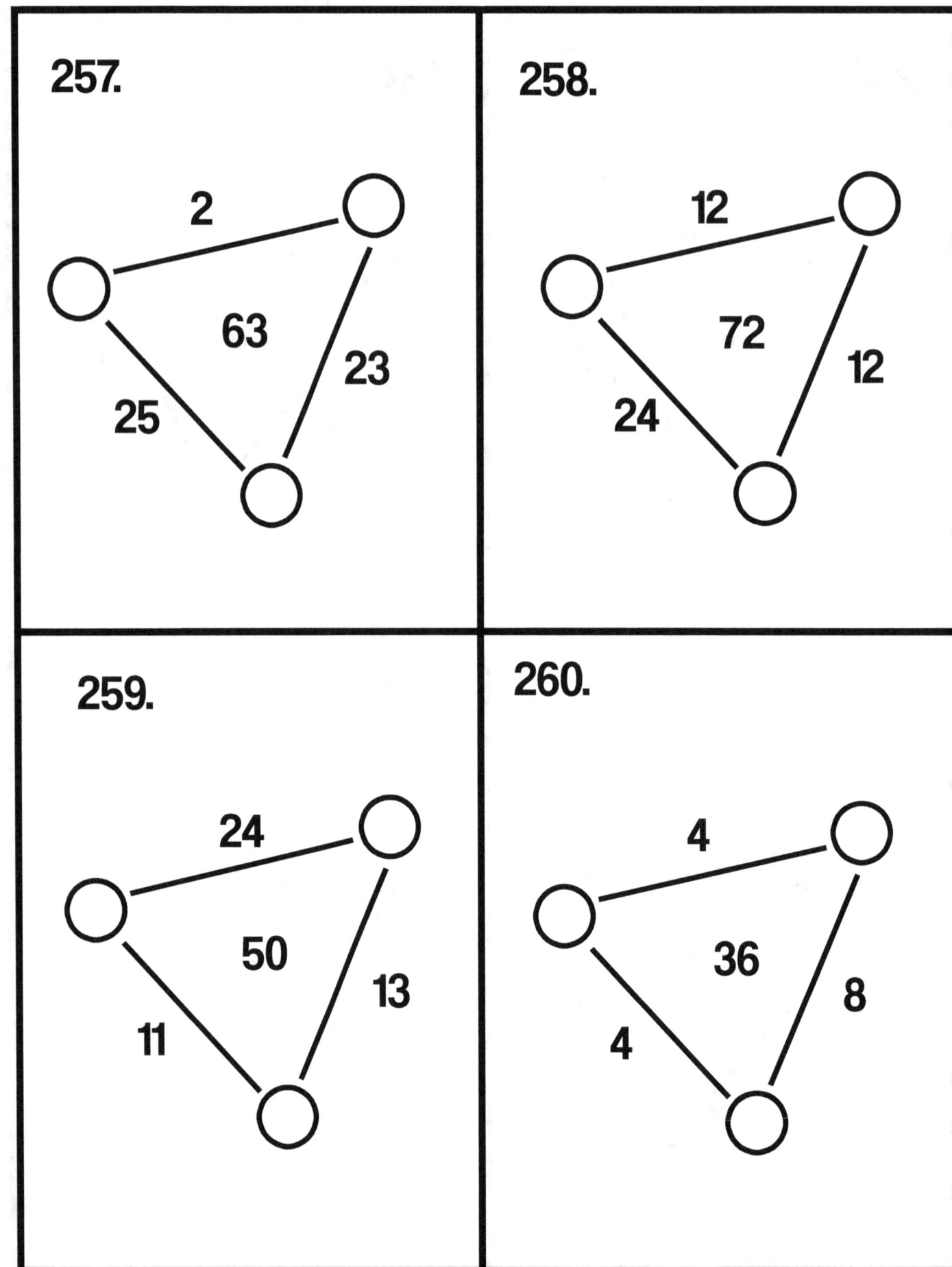

**257.**

2

63

23

25

**258.**

12

72

12

24

**259.**

24

50

13

11

**260.**

4

36

8

4

# Juggle 3 - Subtract

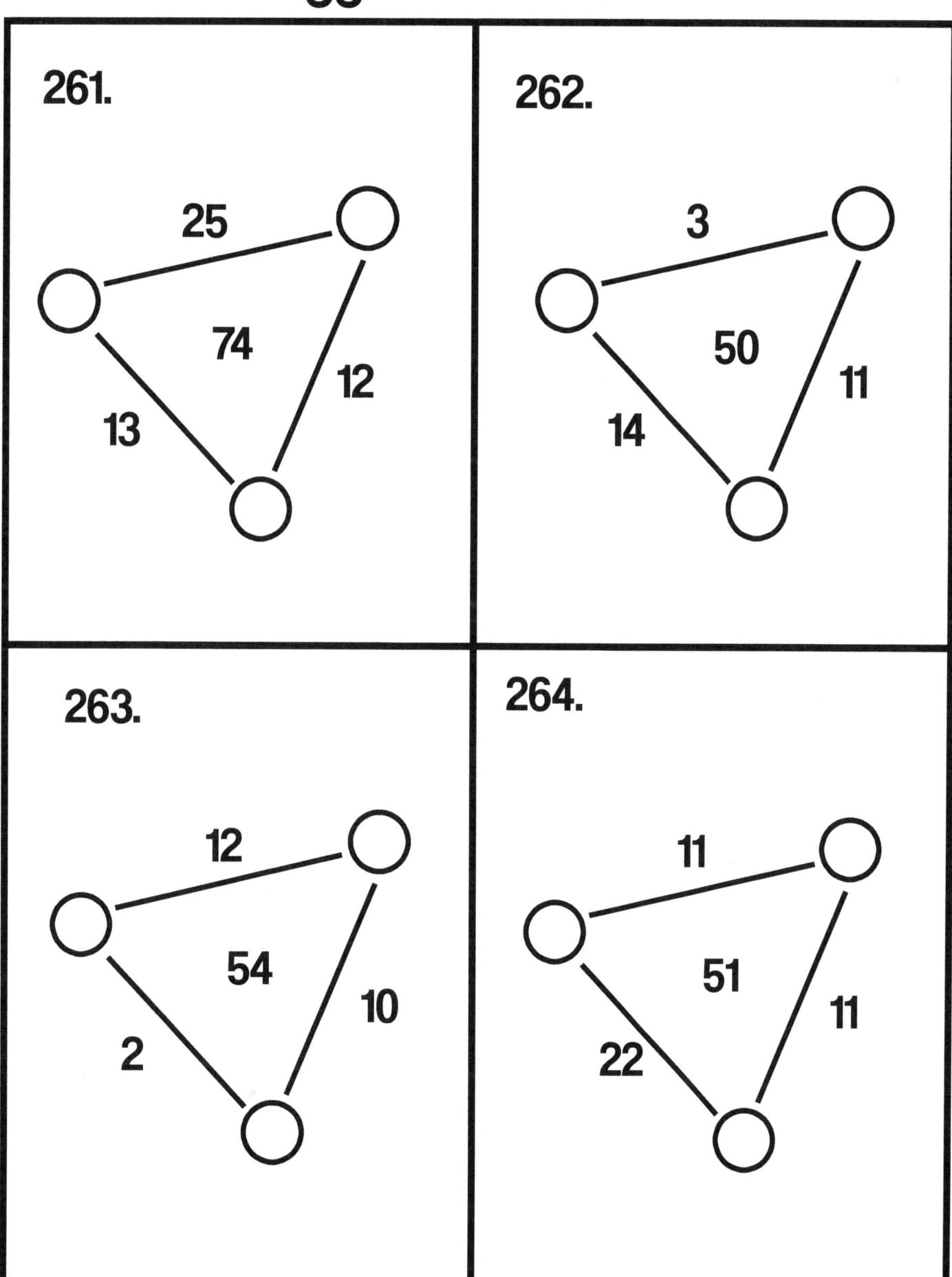

261.

25

74

12

13

262.

3

50

11

14

263.

12

54

10

2

264.

11

51

11

22

# Juggle 3 - Subtract

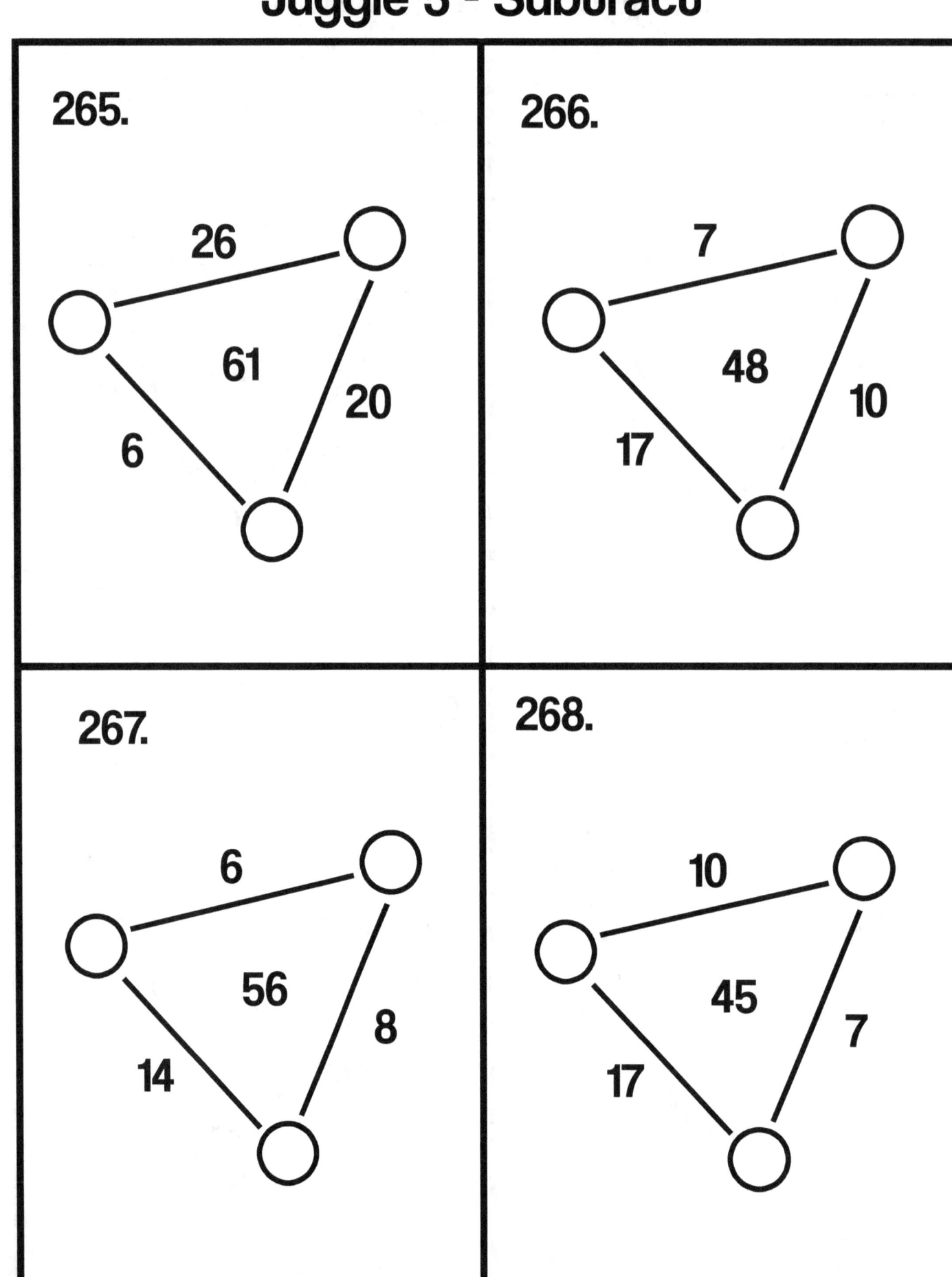

265.

26
61
20
6

266.

7
48
10
17

267.

6
56
8
14

268.

10
45
7
17

# Juggle 3 - Subtract

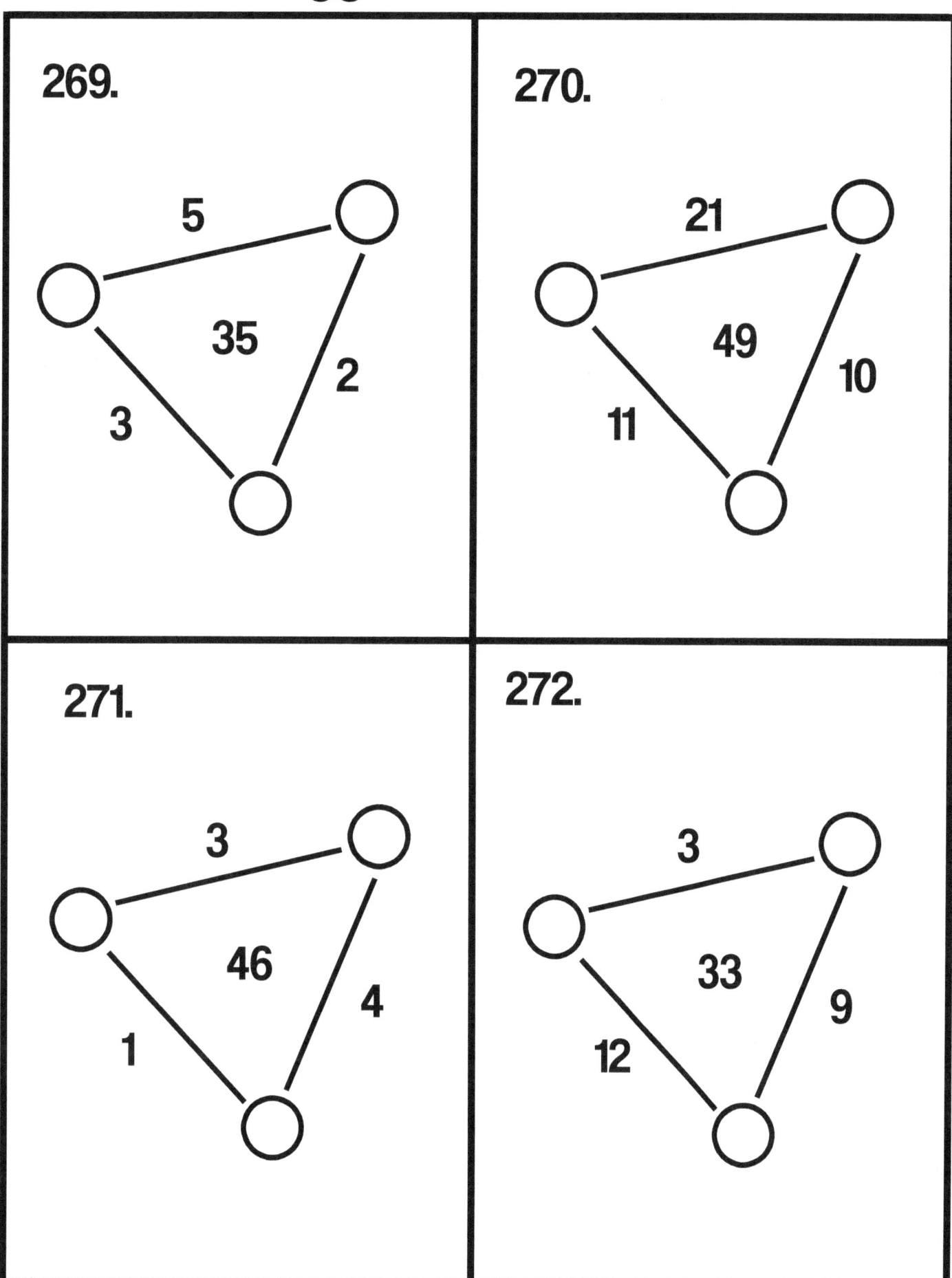

269.

5

35

2

3

270.

21

49

10

11

271.

3

46

4

1

272.

3

33

9

12

# Juggle 3 - Subtract

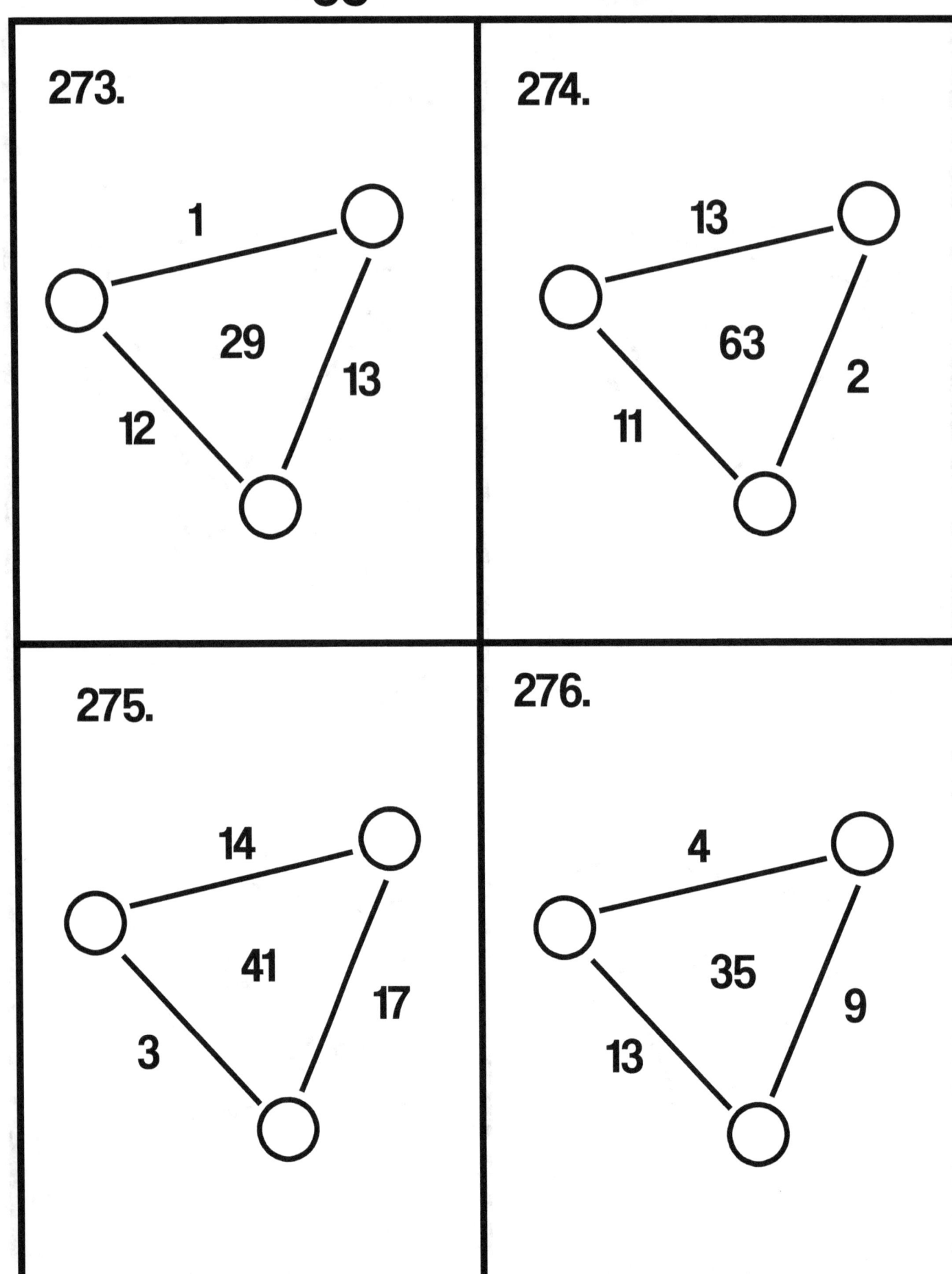

273.

1

29

13

12

274.

13

63

2

11

275.

14

41

17

3

276.

4

35

9

13

# Juggle 3 - Subtract

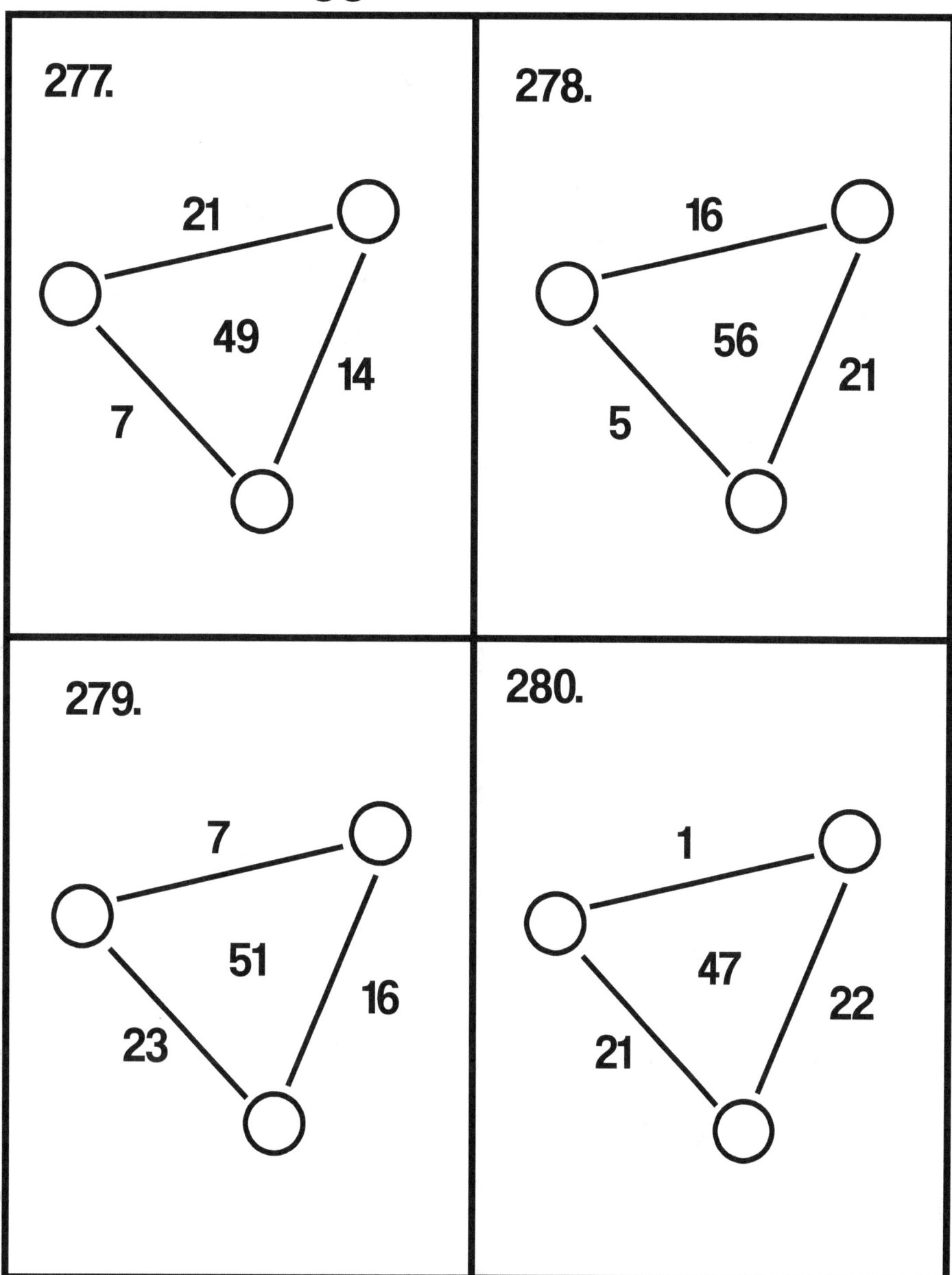

277.

21

49

14

7

278.

16

56

21

5

279.

7

51

16

23

280.

1

47

22

21

# Juggle 3 - Subtract

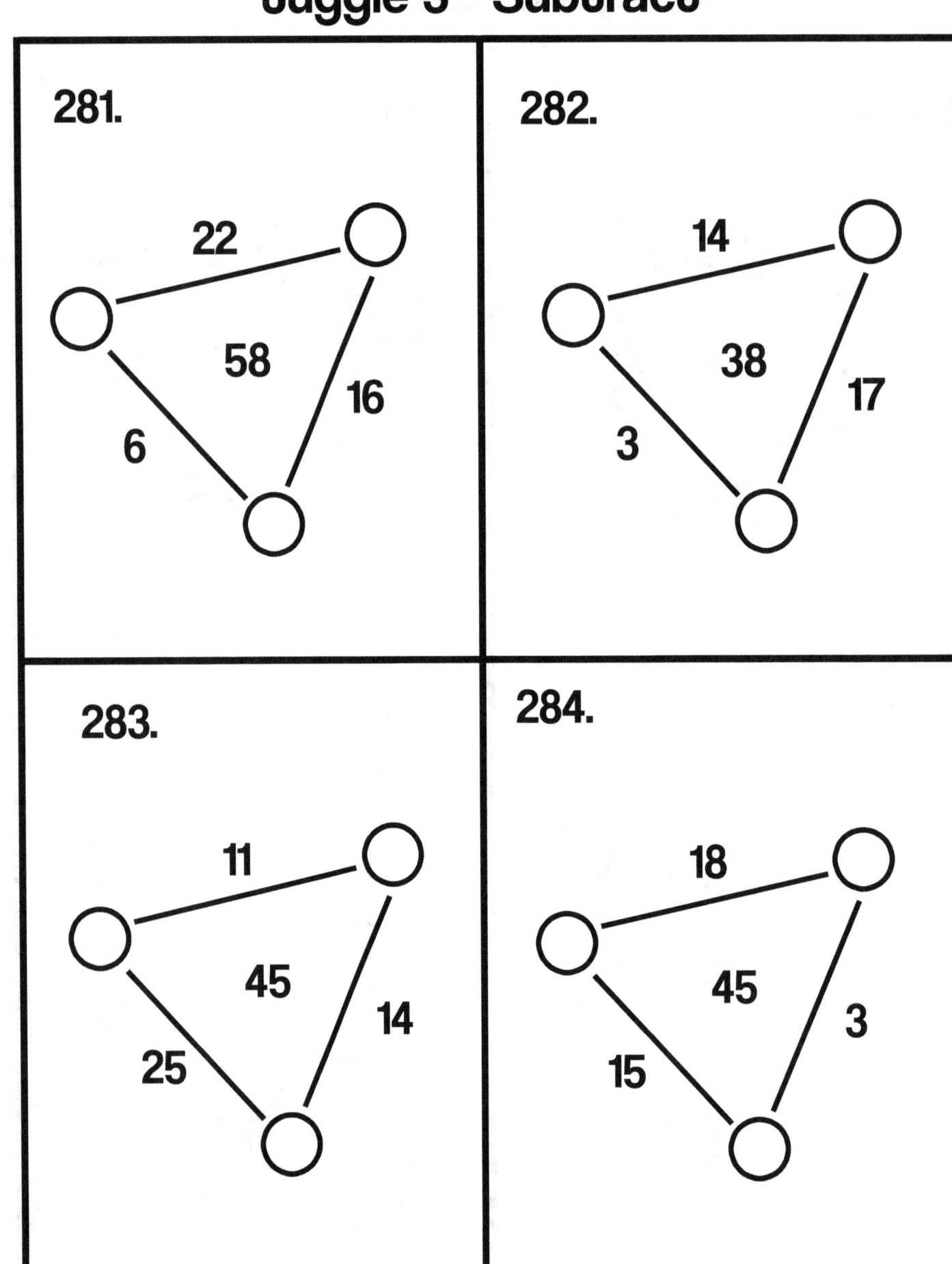

281.

22

58

16

6

282.

14

38

17

3

283.

11

45

14

25

284.

18

45

3

15

# Juggle 3 - Subtract

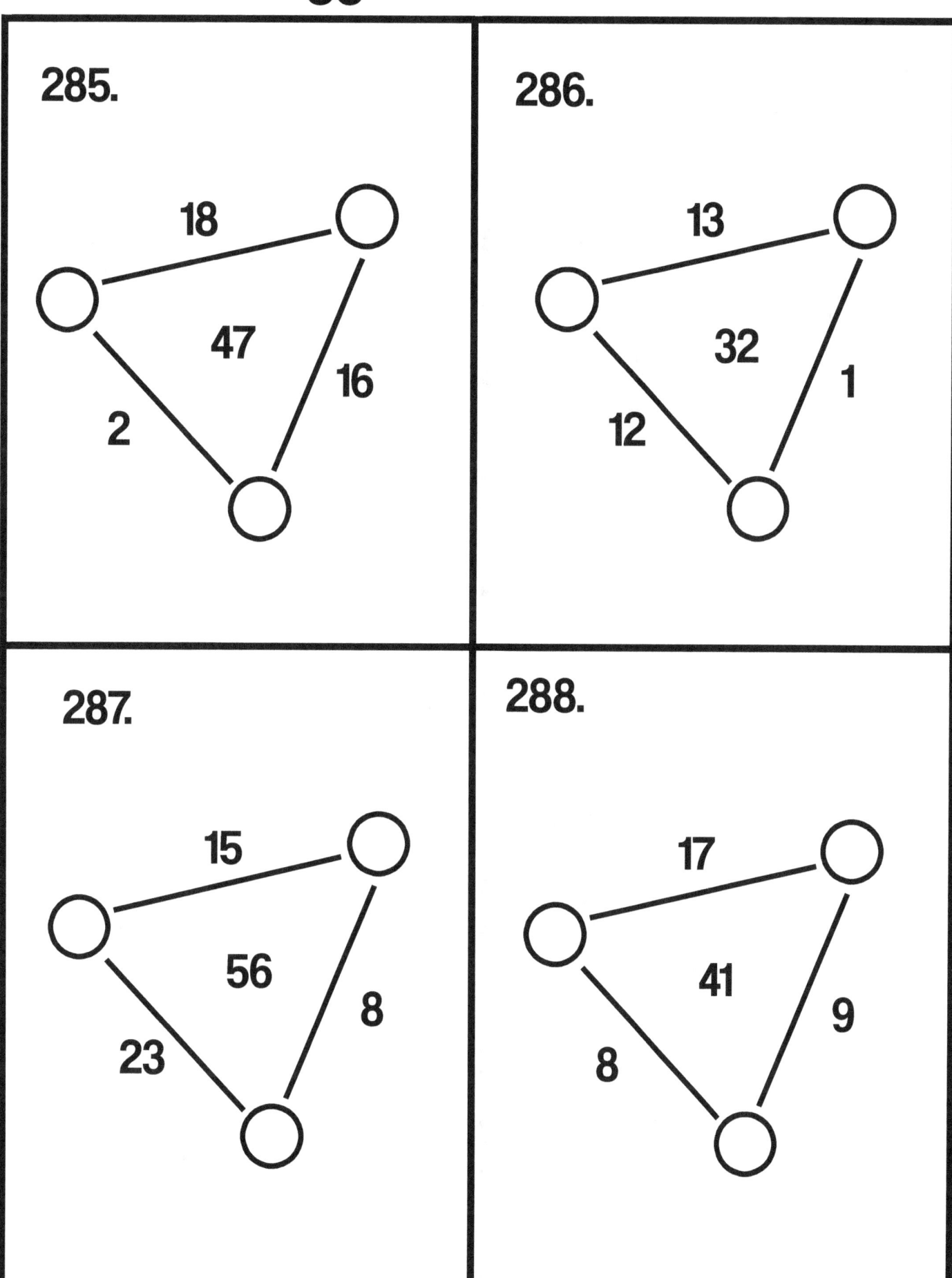

285.

18

47

16

2

286.

13

32

1

12

287.

15

56

8

23

288.

17

41

9

8

# Juggle 3 - Subtract

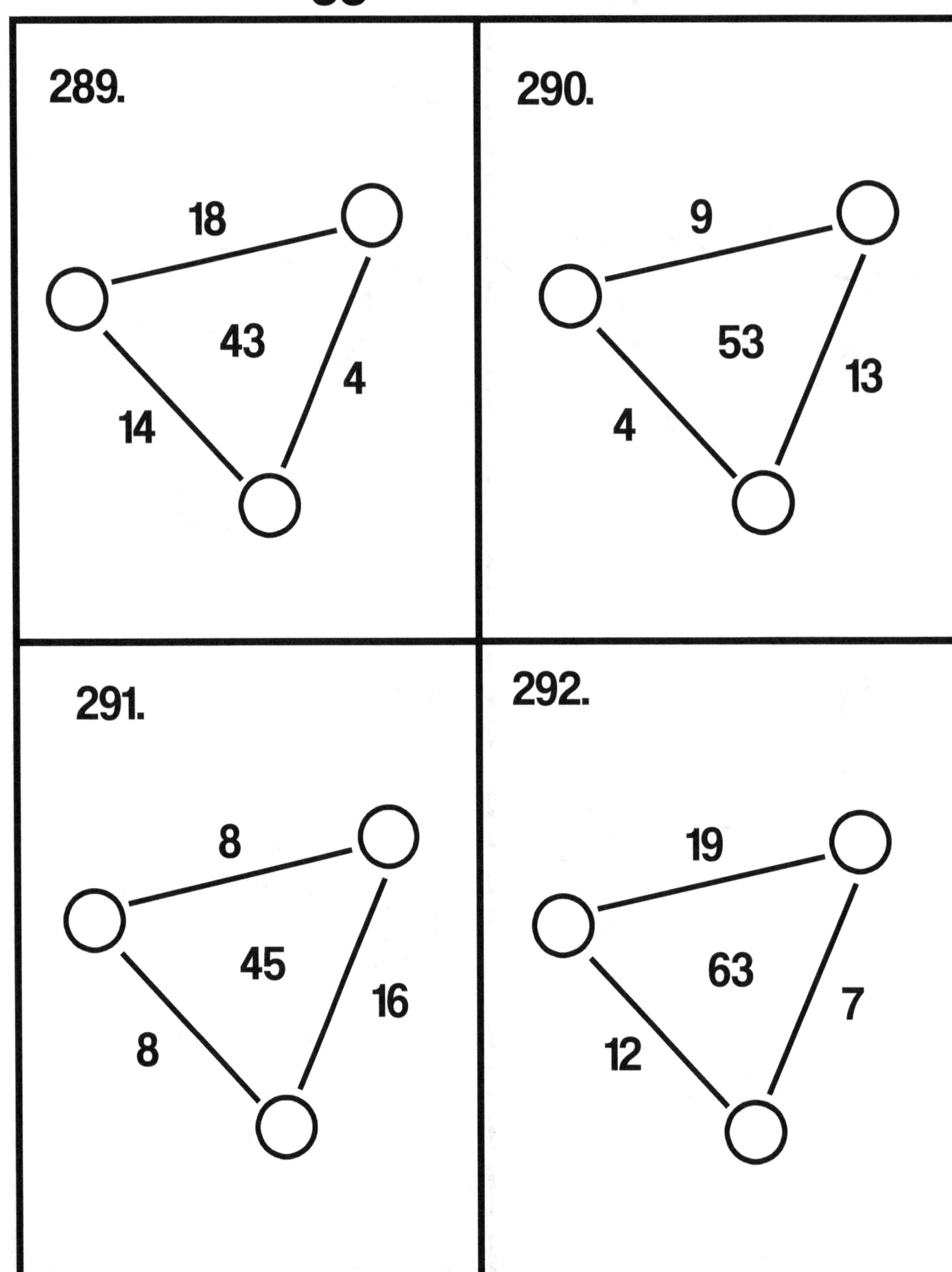

289.

290.

291.

292.

# Juggle 3 - Subtract

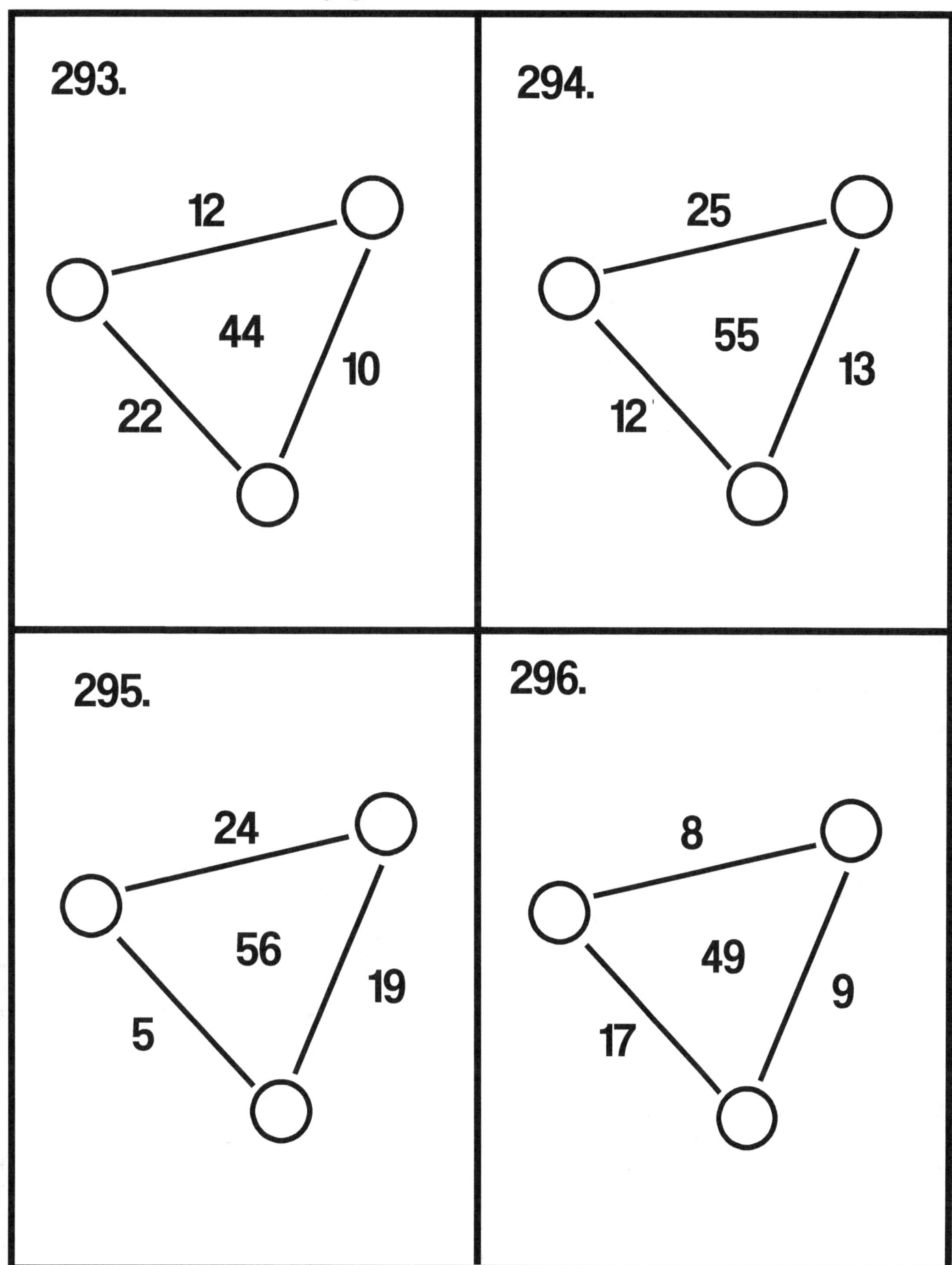

293.

12

44

10

22

294.

25

55

13

12

295.

24

56

19

5

296.

8

49

9

17

# Juggle 3 - Subtract

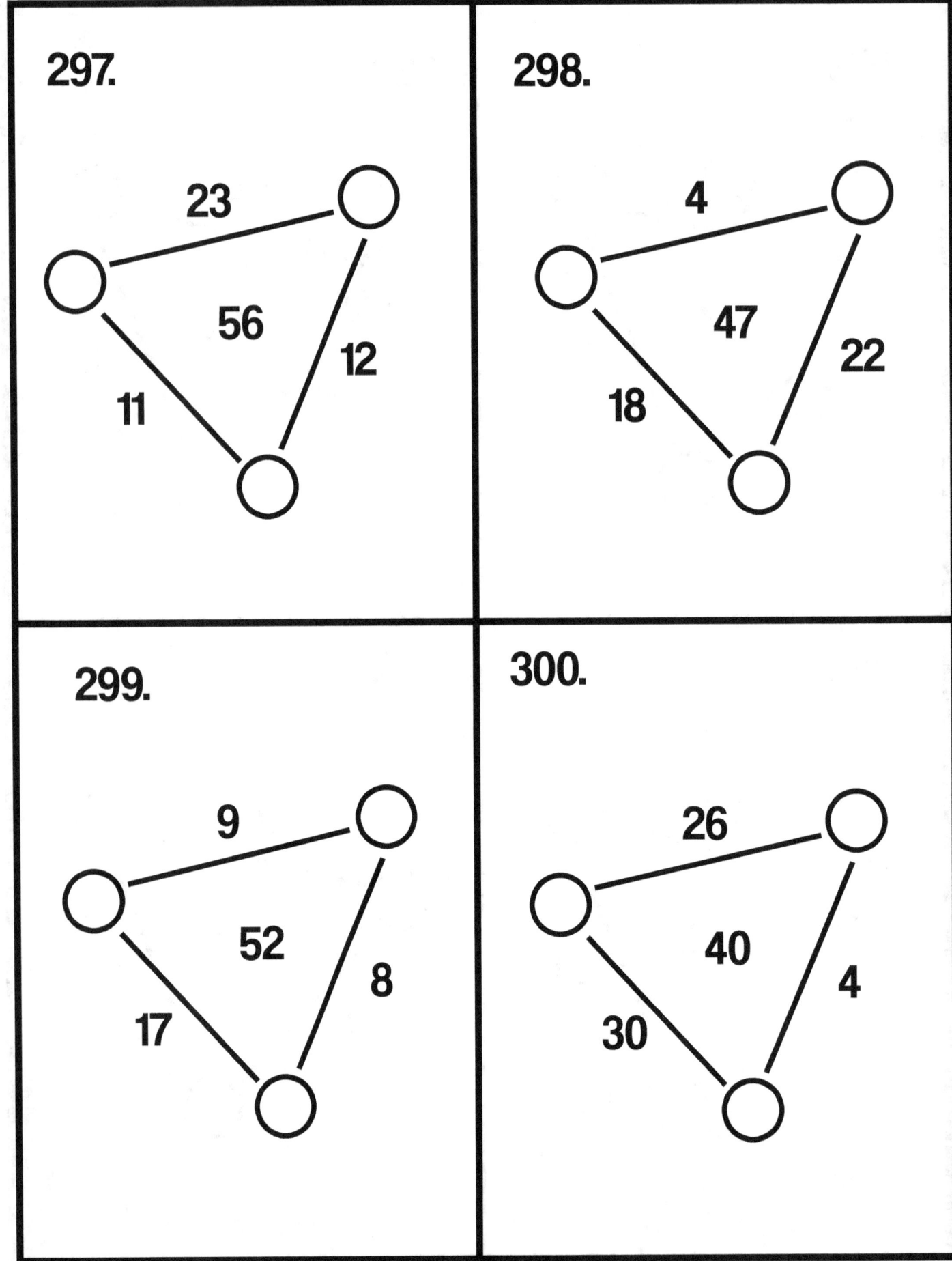

297.
23
56
12
11

298.
4
47
22
18

299.
9
52
8
17

300.
26
40
4
30

# Juggle 3 - Subtract

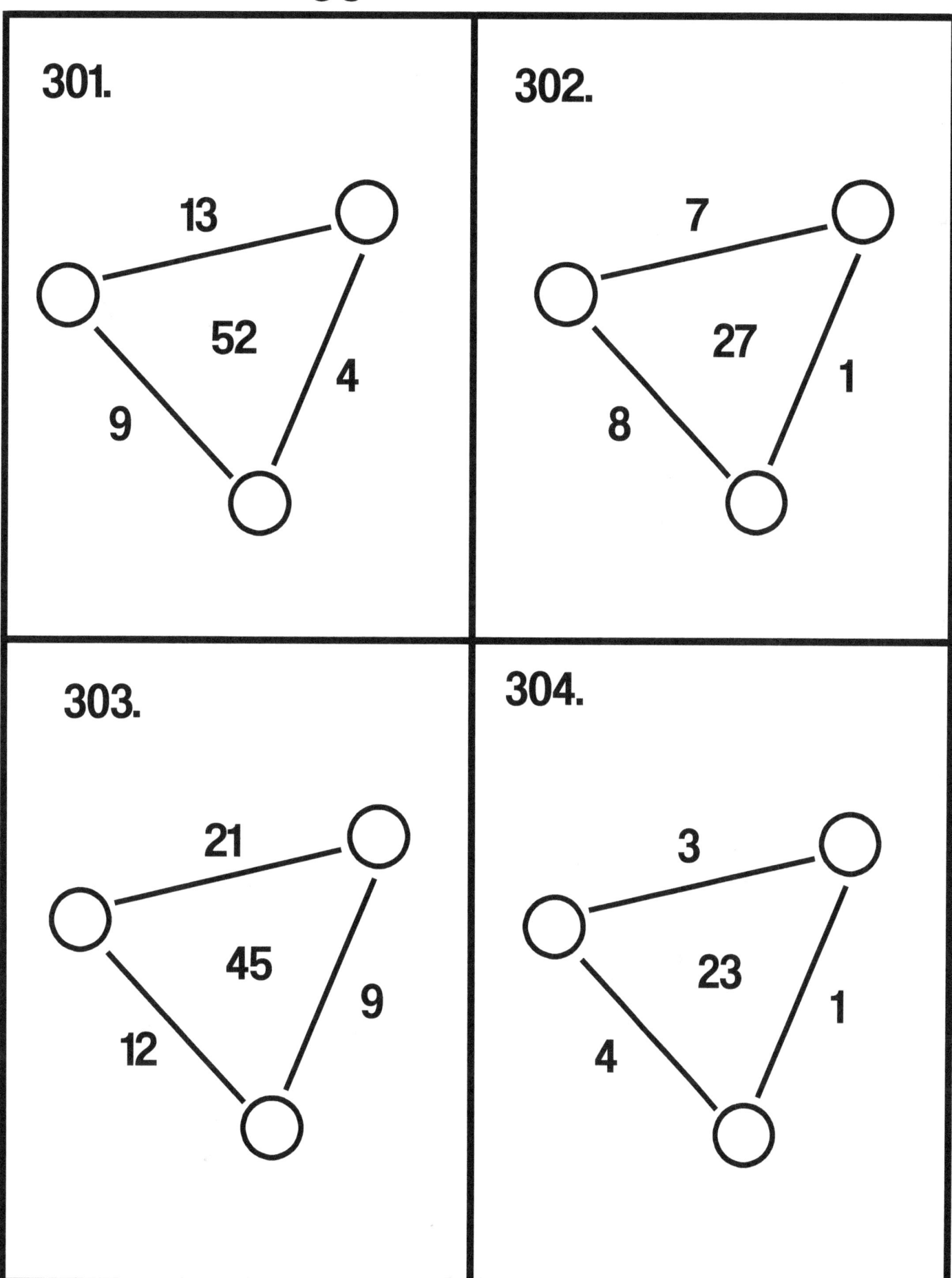

301.

13
52
4
9

302.

7
27
1
8

303.

21
45
9
12

304.

3
23
1
4

# Juggle 3 - Subtract

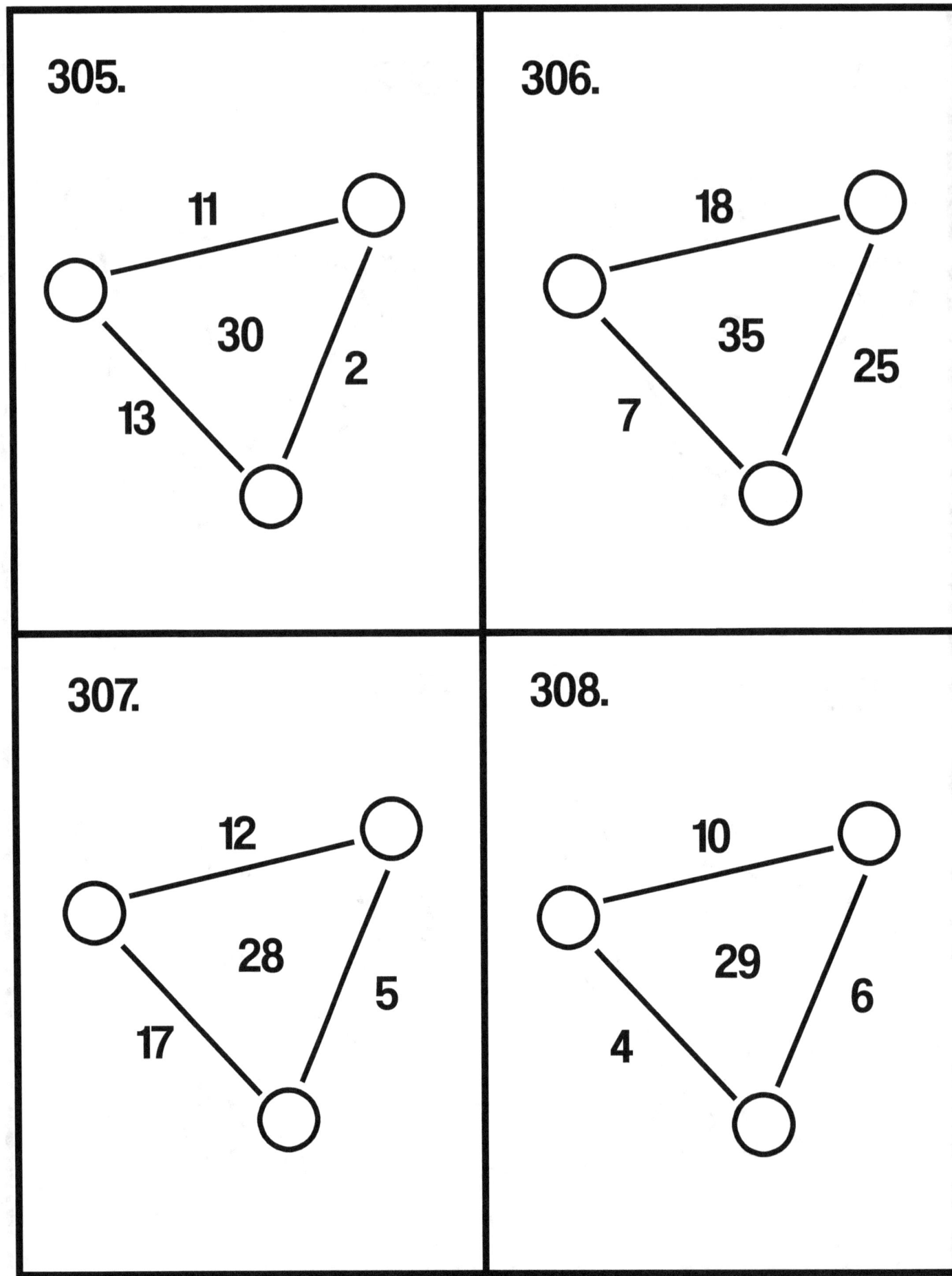

305.

11

30

13

2

306.

18

35

7

25

307.

12

28

17

5

308.

10

29

4

6

# Juggle 3 - Subtract

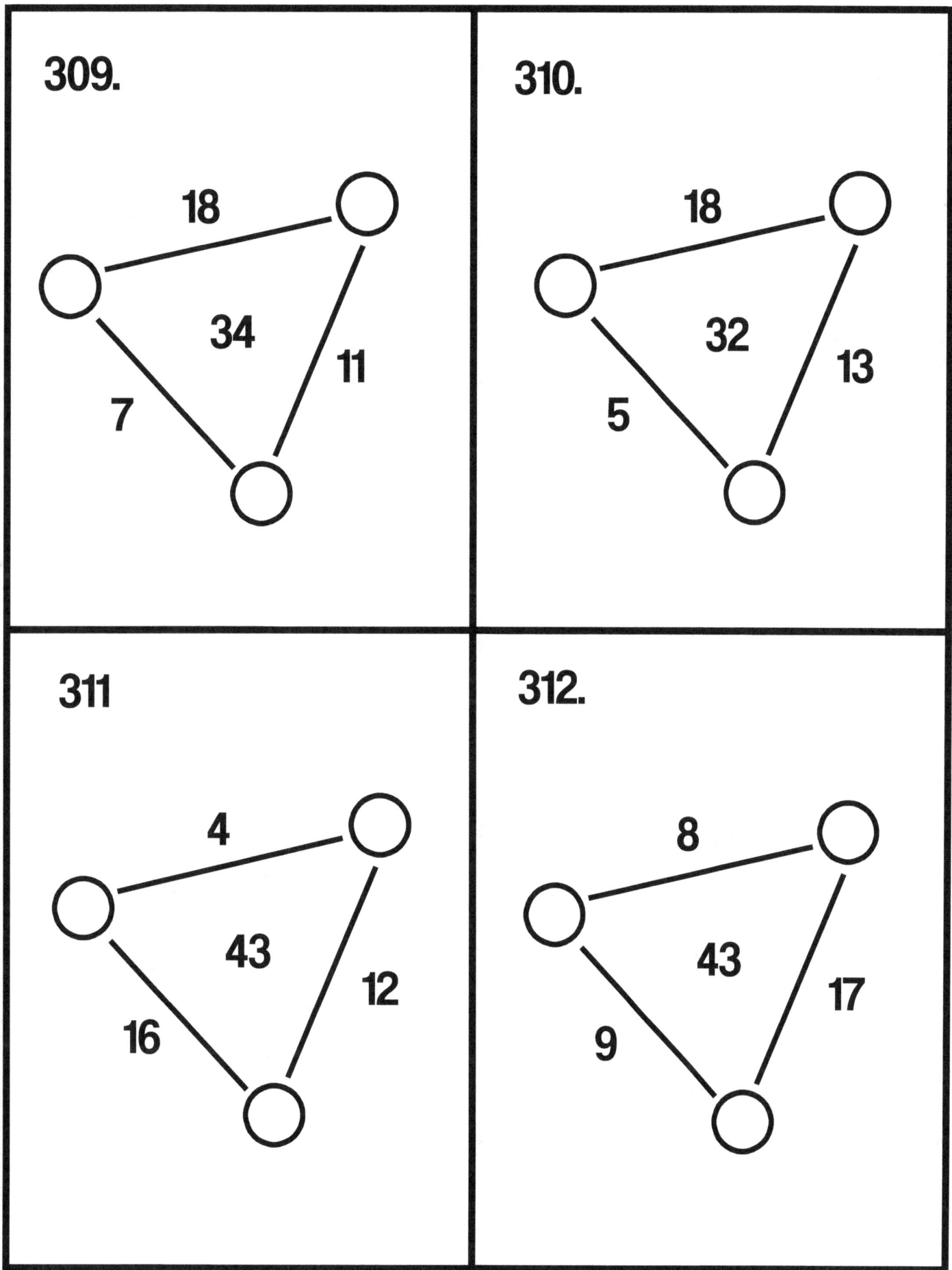

# Juggle 3 - Subtract

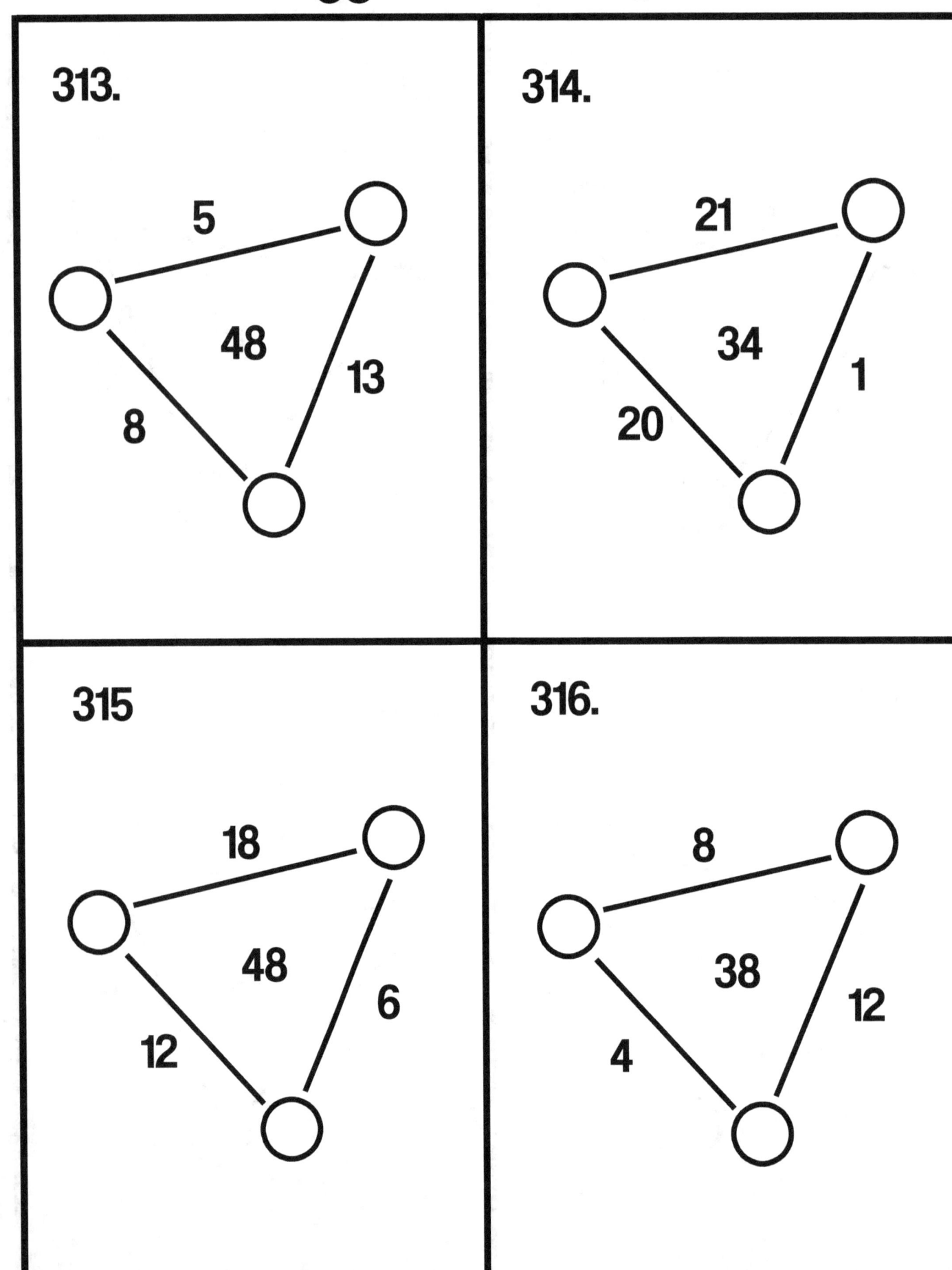

313.

5
48
13
8

314.

21
34
1
20

315

18
48
6
12

316.

8
38
12
4

# Juggle 3 - Subtract

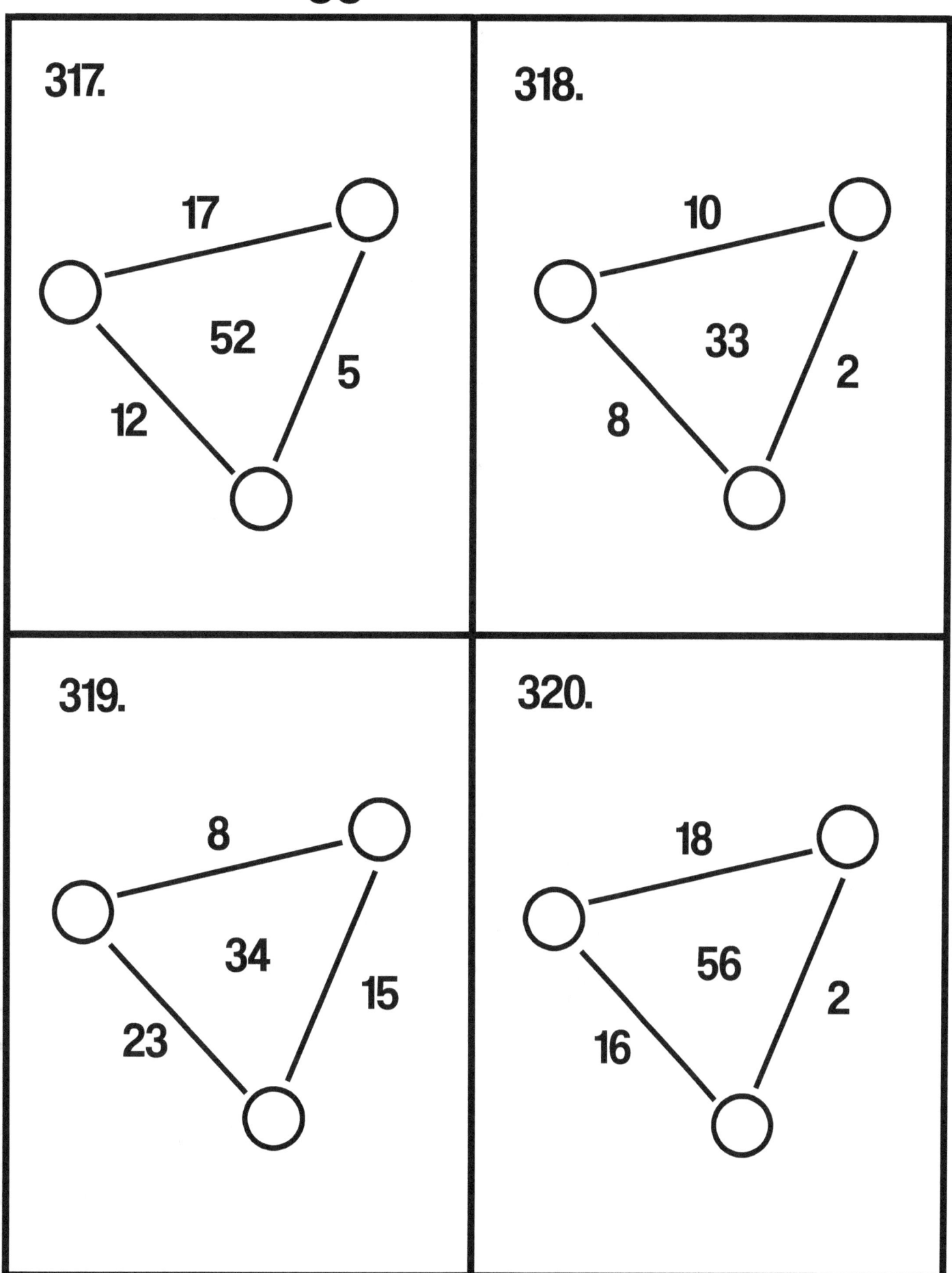

317.

17

52

5

12

318.

10

33

2

8

319.

8

34

15

23

320.

18

56

2

16

# Juggle 3 - Subtract

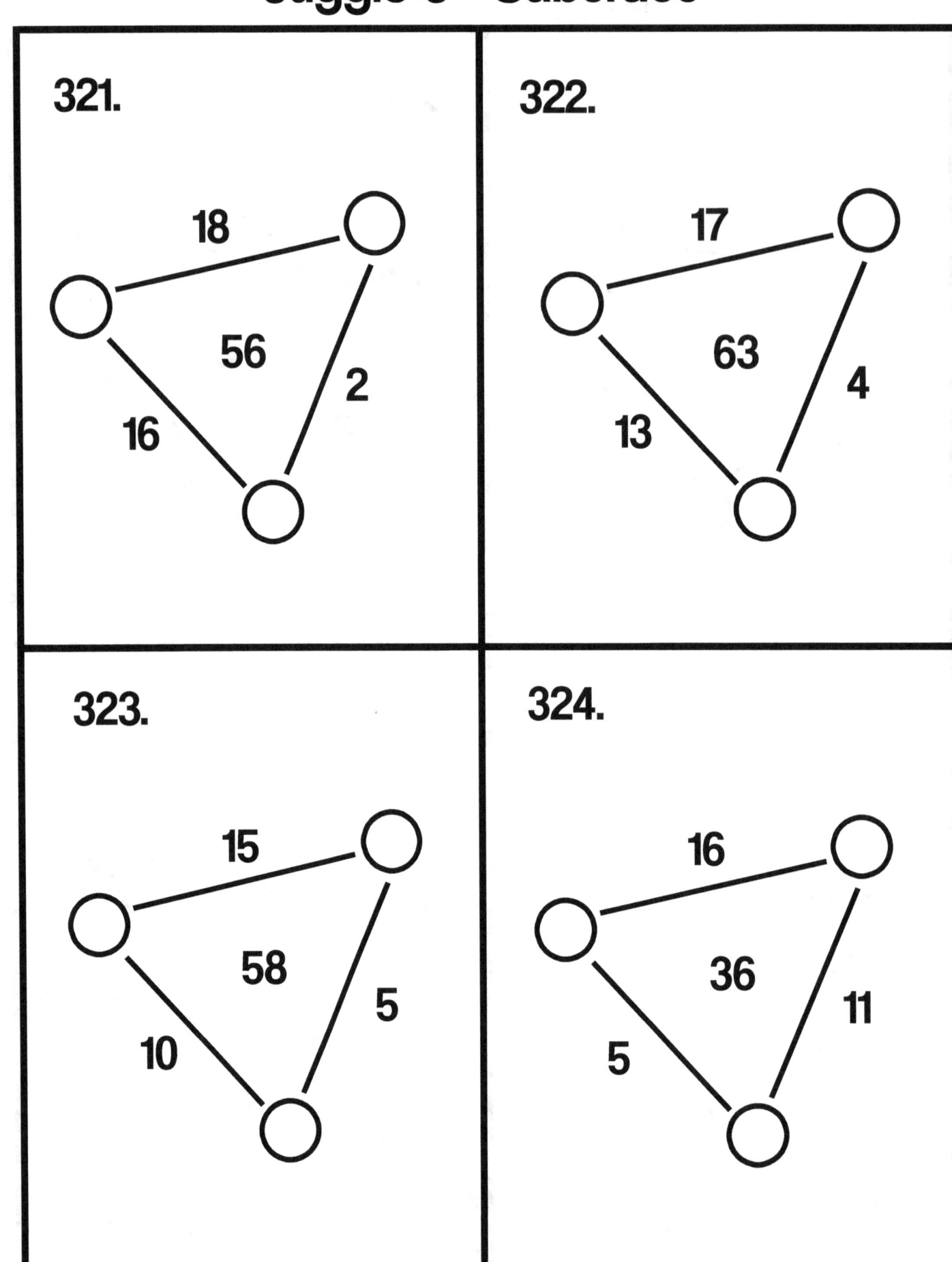

321.

18

56

2

16

322.

17

63

4

13

323.

15

58

5

10

324.

16

36

11

5

# Juggle 3 - Subtract

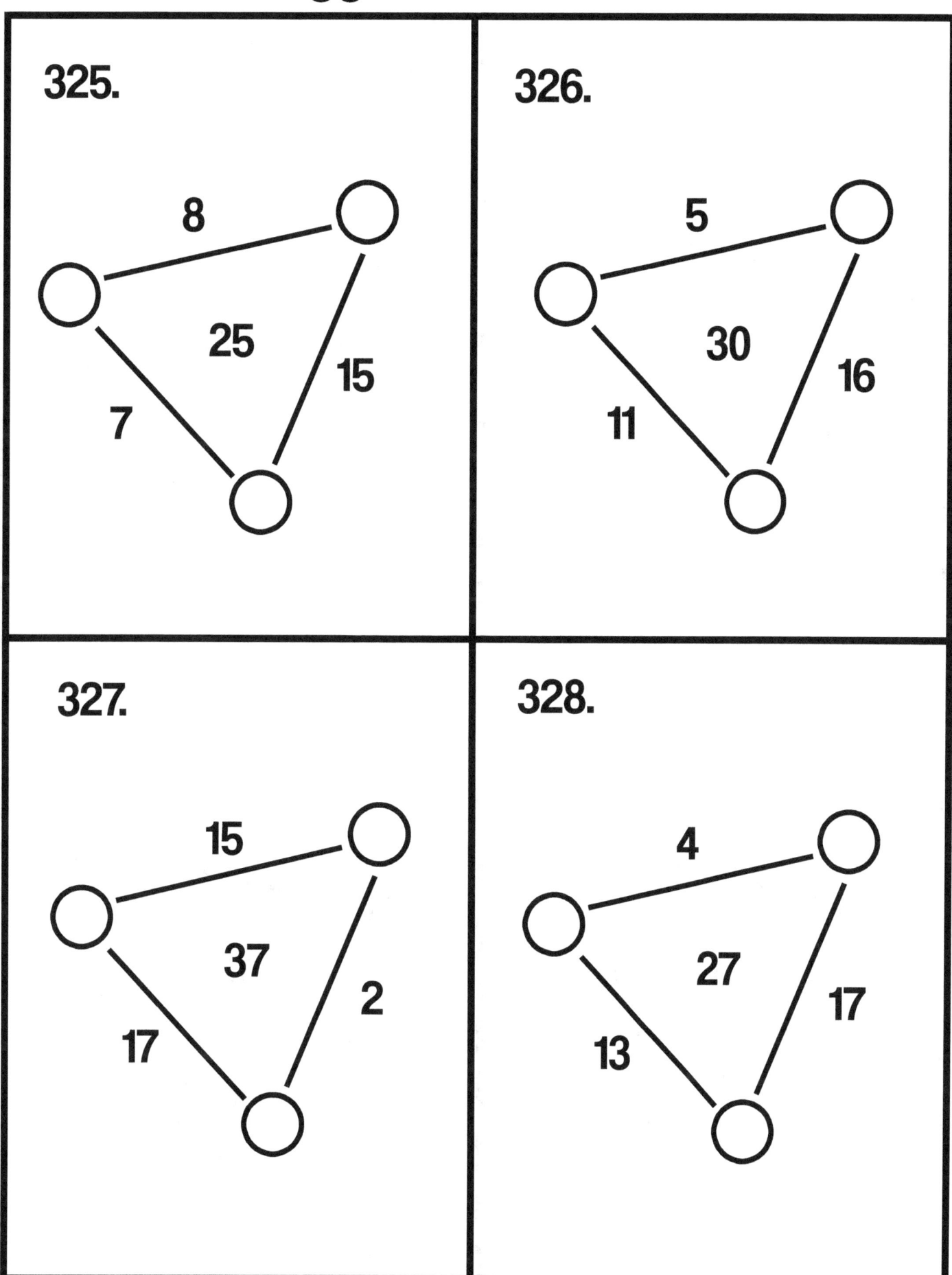

325.

8
25
15
7

326.

5
30
16
11

327.

15
37
2
17

328.

4
27
17
13

# Juggle 3 - Subtract

**329.**

**330.**

**331.**

**332.**

# Juggle 3 - Subtract

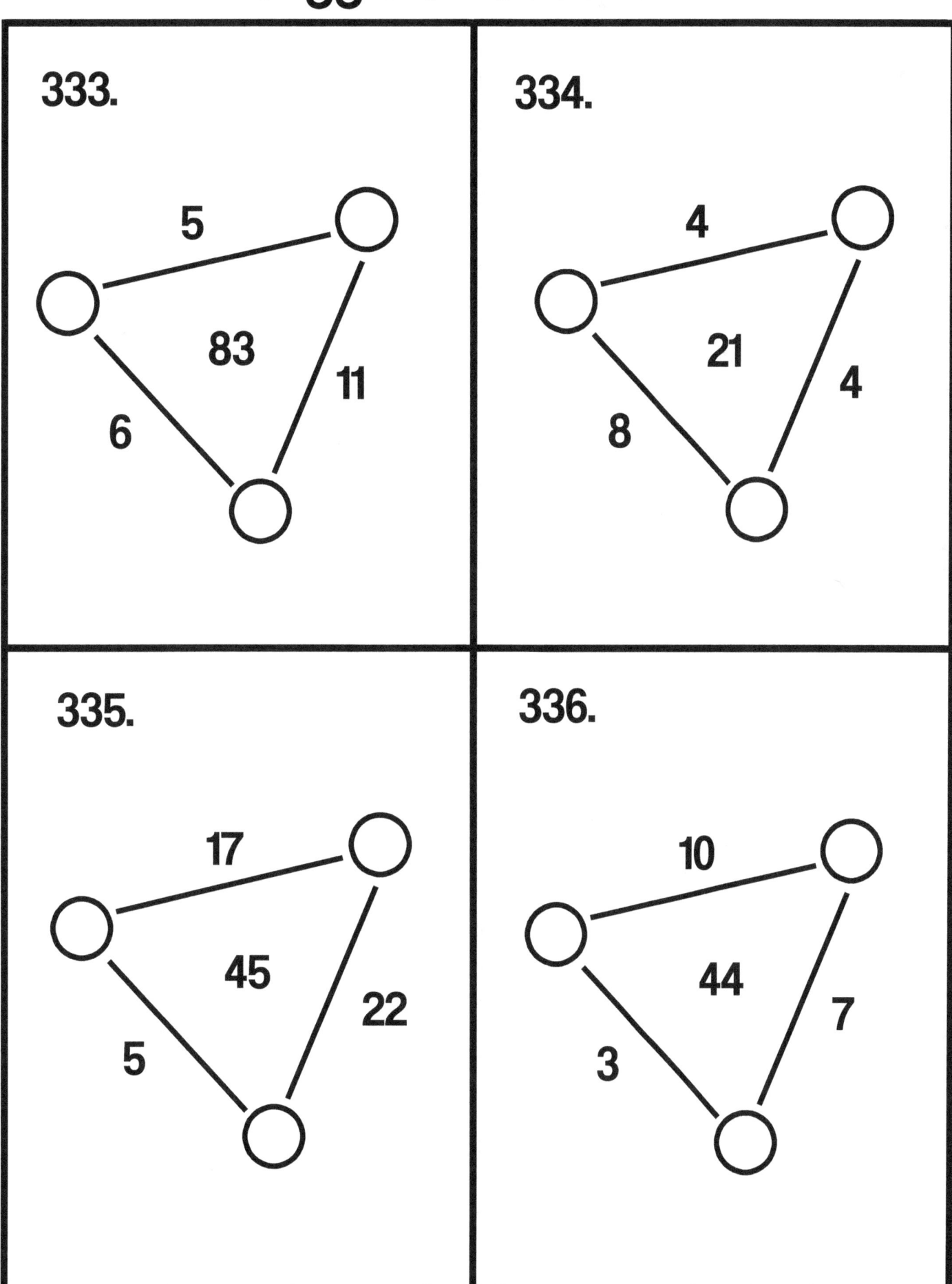

**333.**

5

83

11

6

**334.**

4

21

4

8

**335.**

17

45

22

5

**336.**

10

44

7

3

# Juggle 3 - Subtract

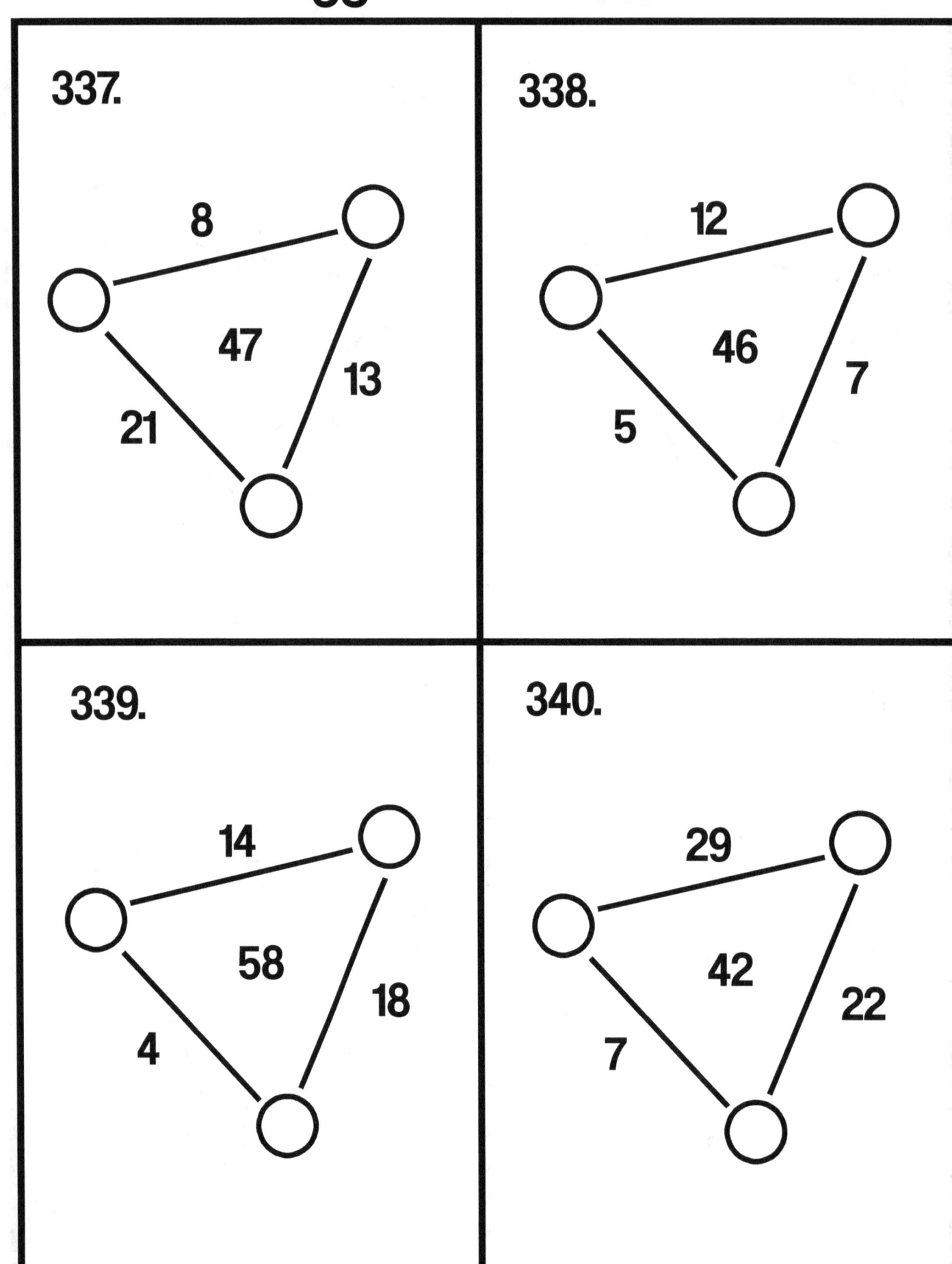

337.

8

47

13

21

338.

12

46

7

5

339.

14

58

18

4

340.

29

42

22

7

# Juggle 3 - Subtract

341.

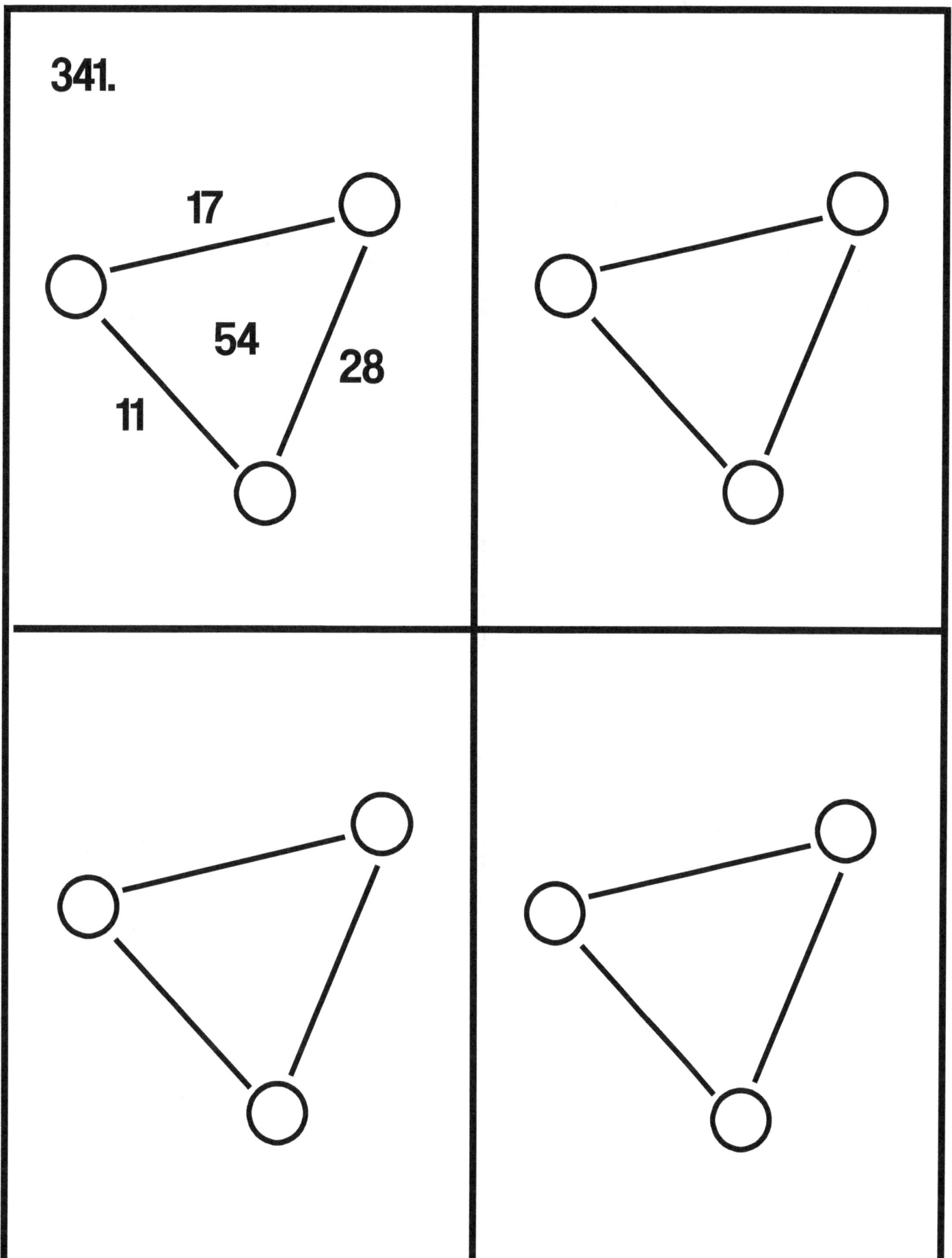

# Juggle 3 - Multiply

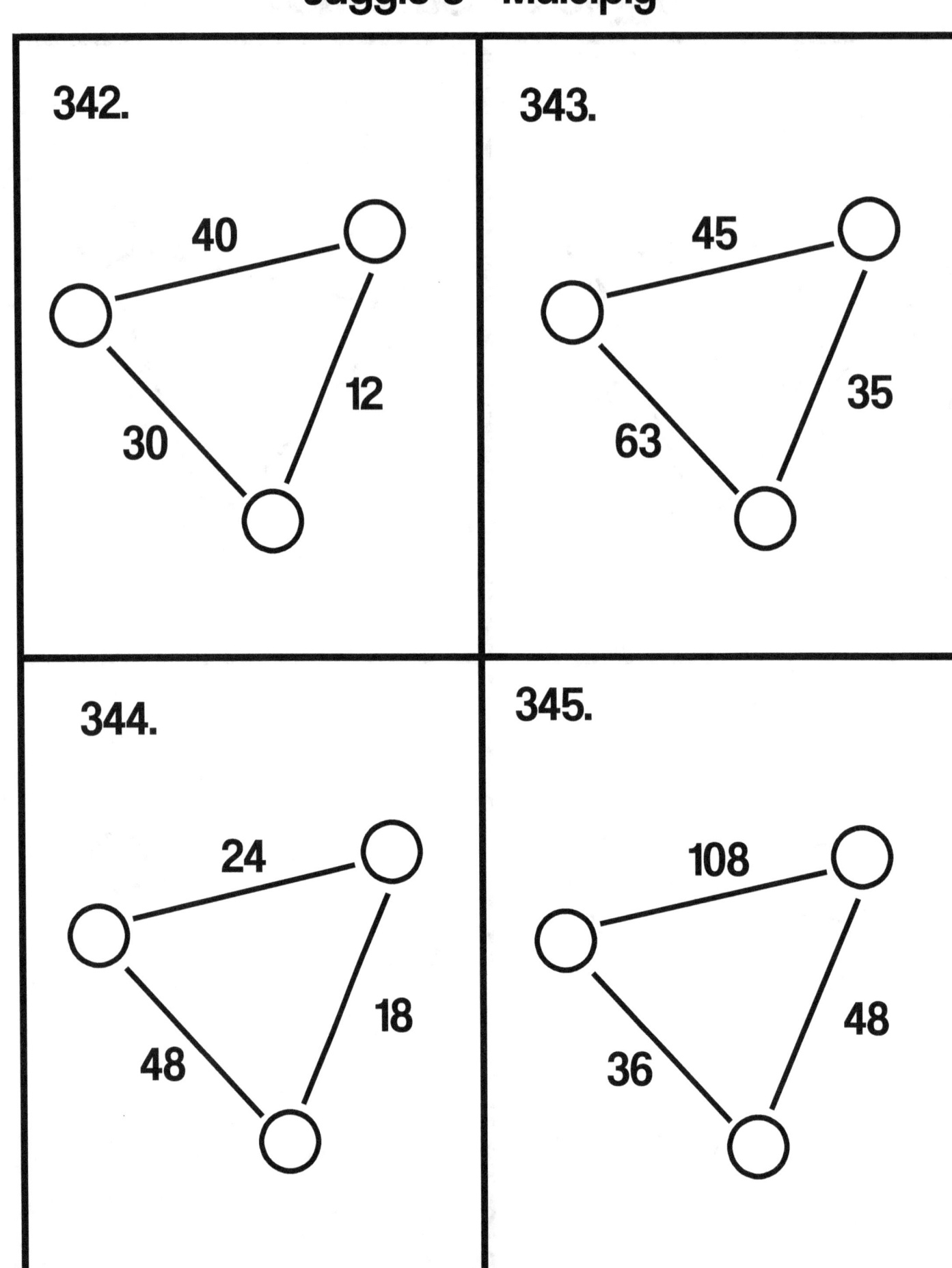

342.

40  30  12

343.

45  63  35

344.

24  48  18

345.

108  36  48

# Juggle 3 - Multiply

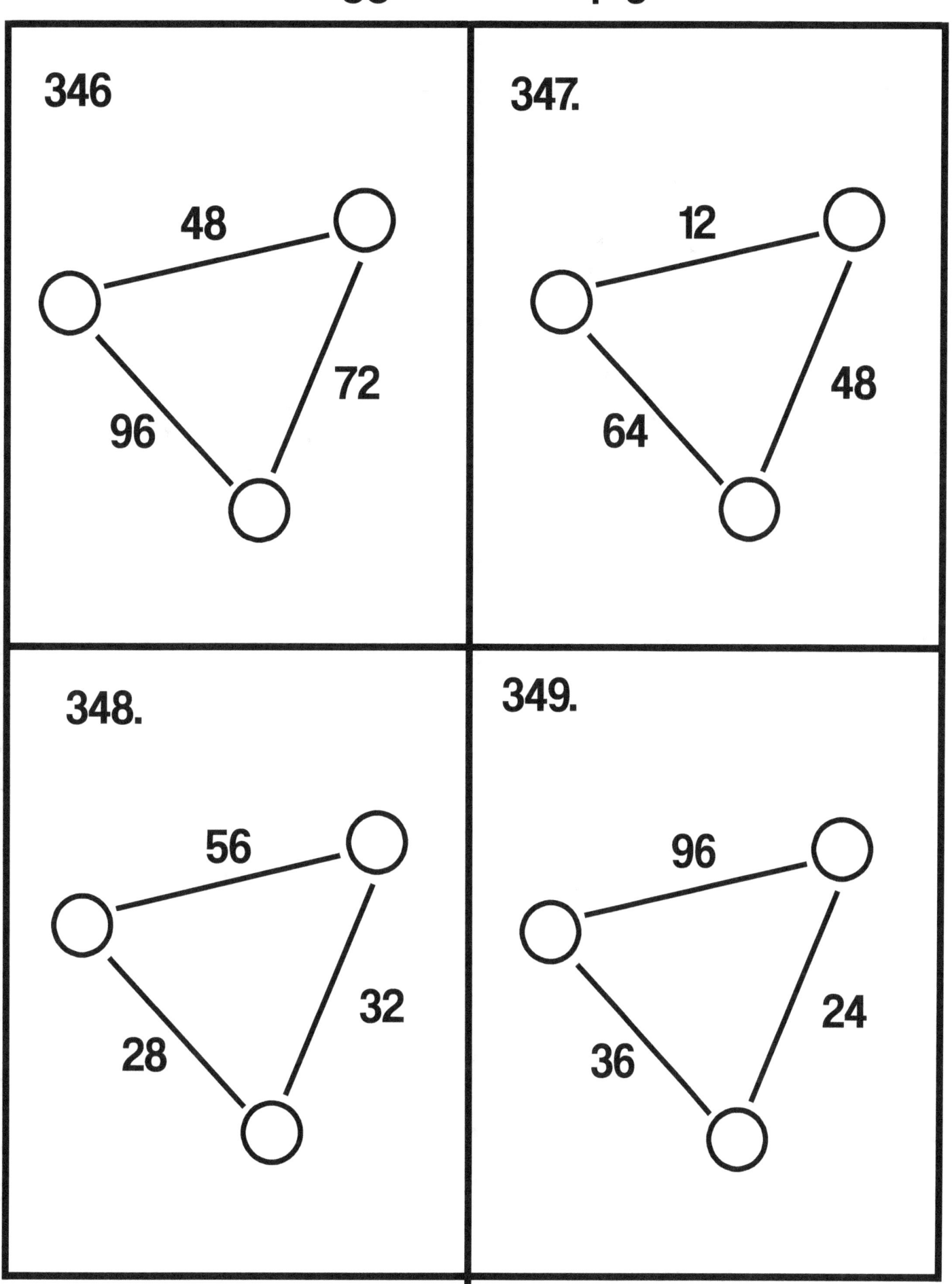

# Juggle 3 - Multiply

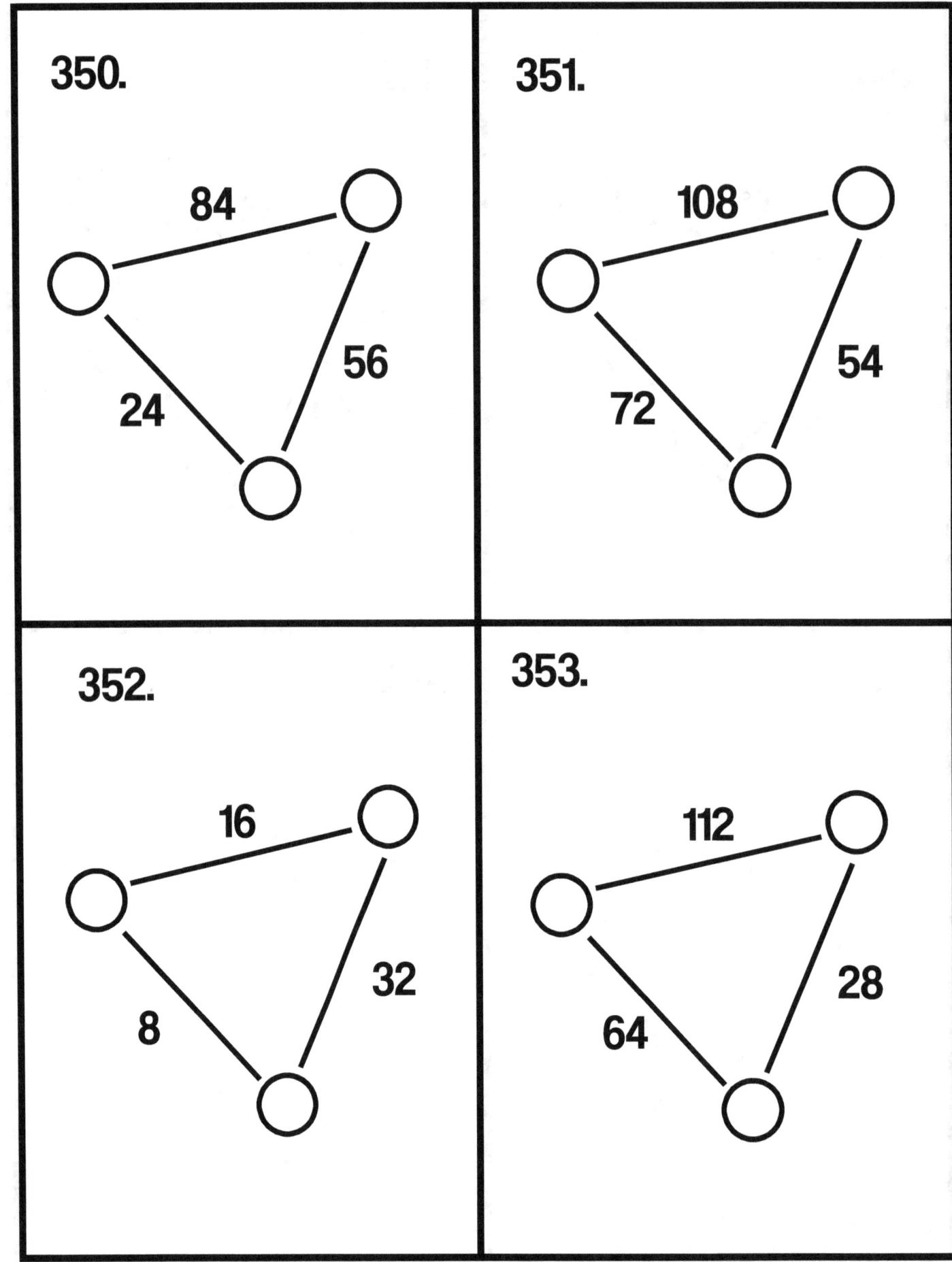

**350.**

84

56

24

**351.**

108

54

72

**352.**

16

32

8

**353.**

112

28

64

# Juggle 3 - Multiply

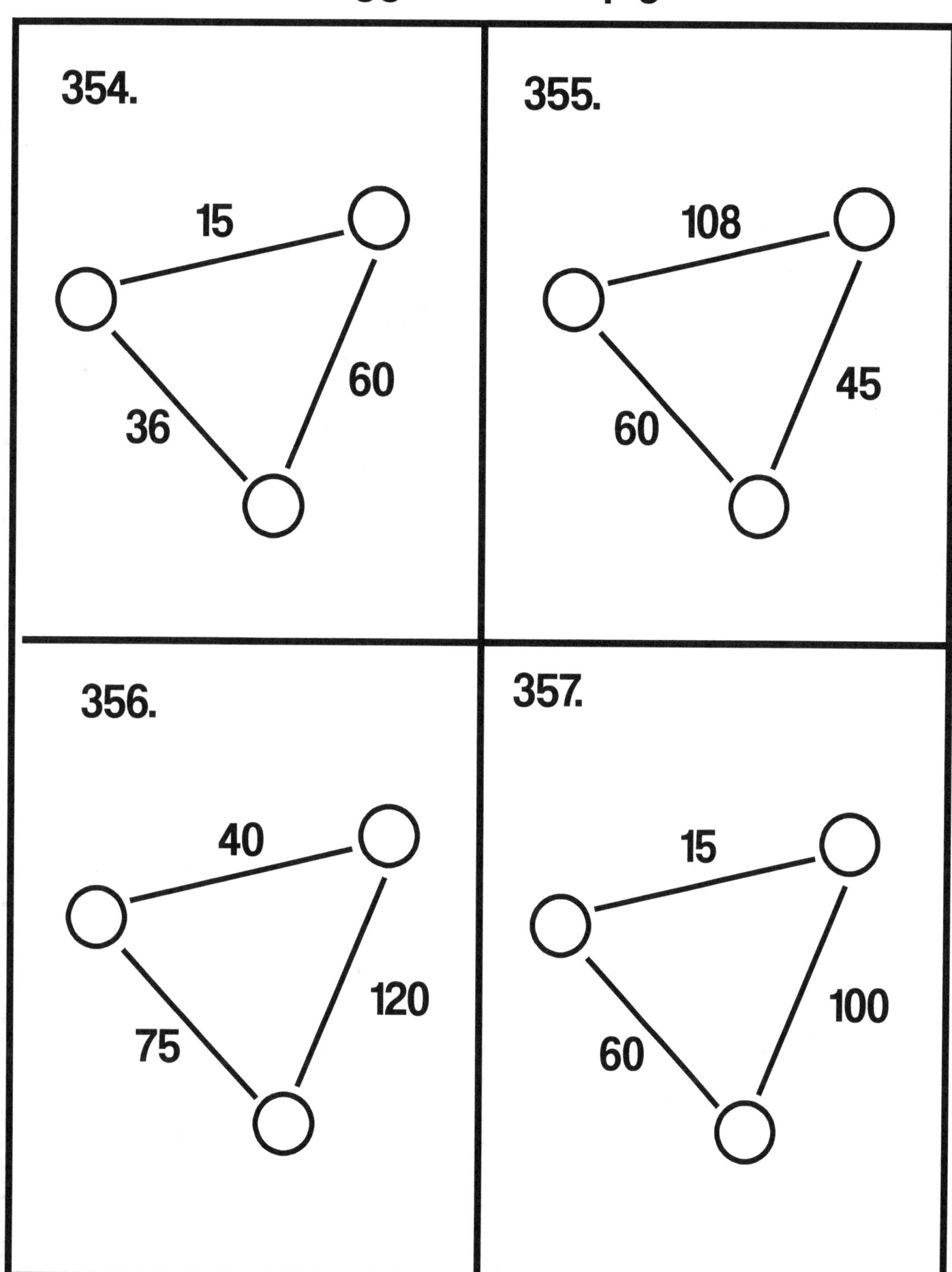

354.

15
60
36

355.

108
45
60

356.

40
120
75

357.

15
100
60

# Juggle 3 - Multiply

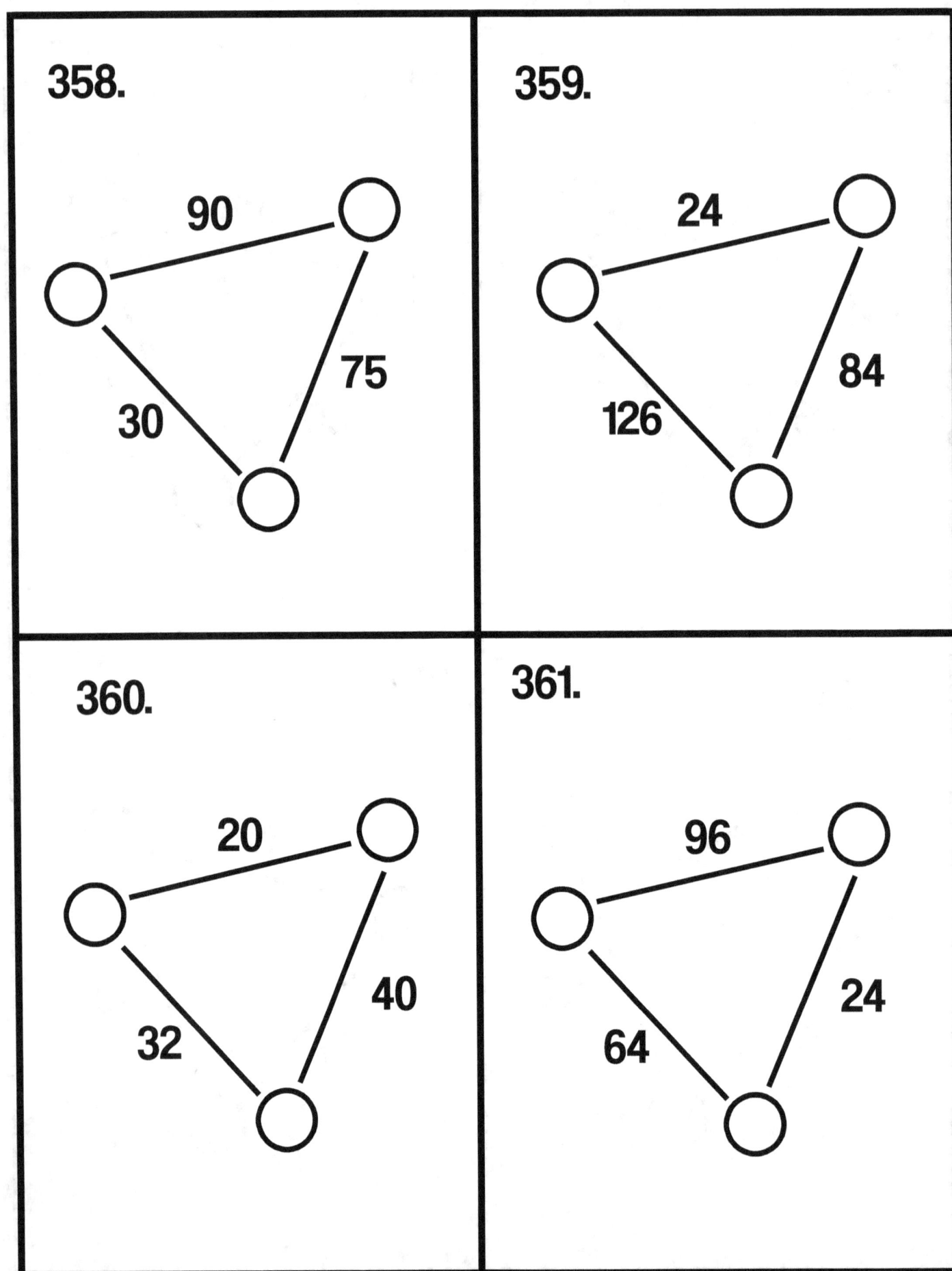

**358.**

90  75  30

**359.**

24  84  126

**360.**

20  40  32

**361.**

96  24  64

# Juggle 3 - Multiply

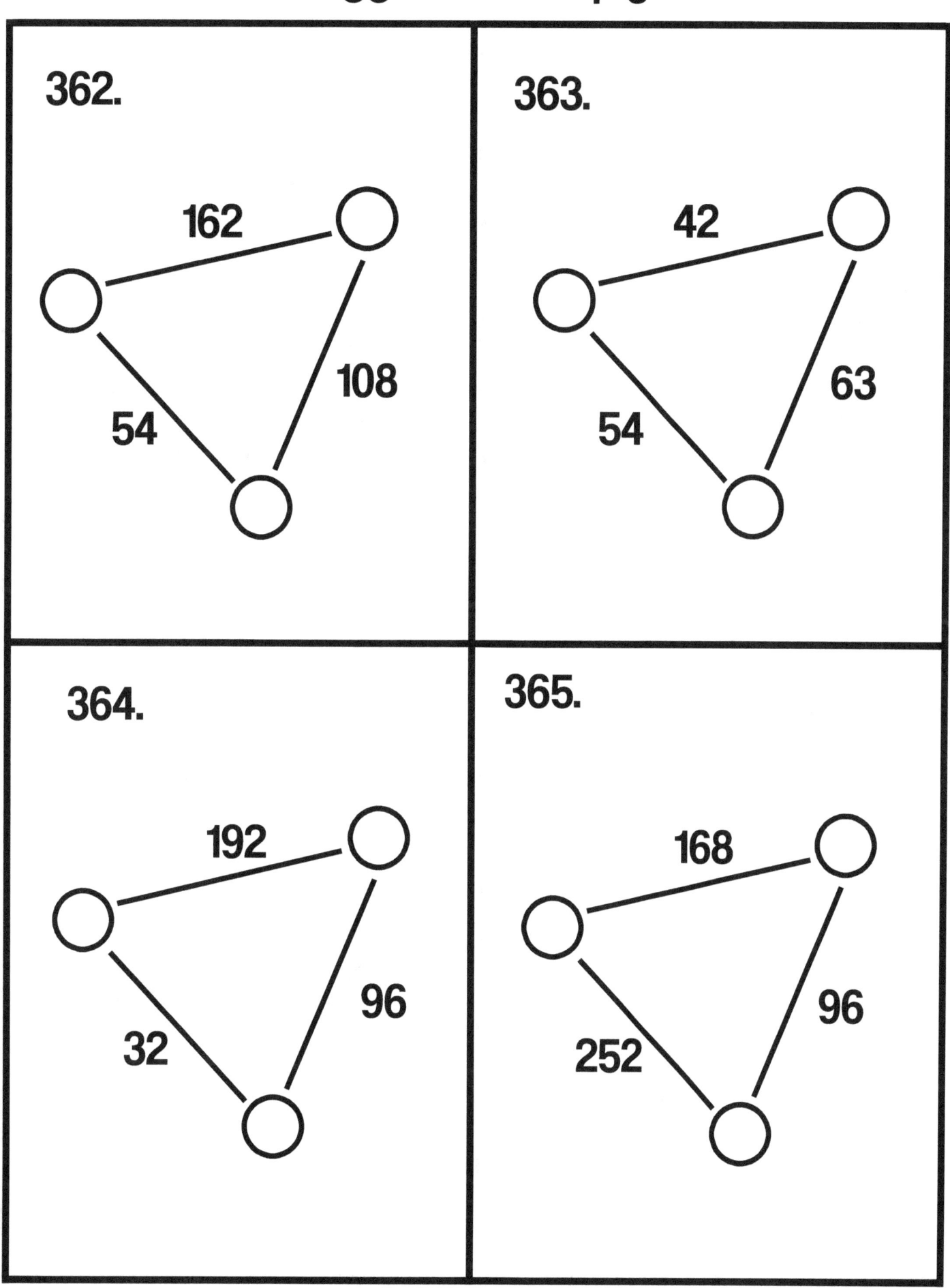

362.

162
108
54

363.

42
63
54

364.

192
96
32

365.

168
96
252

# Juggle 3 - Multiply

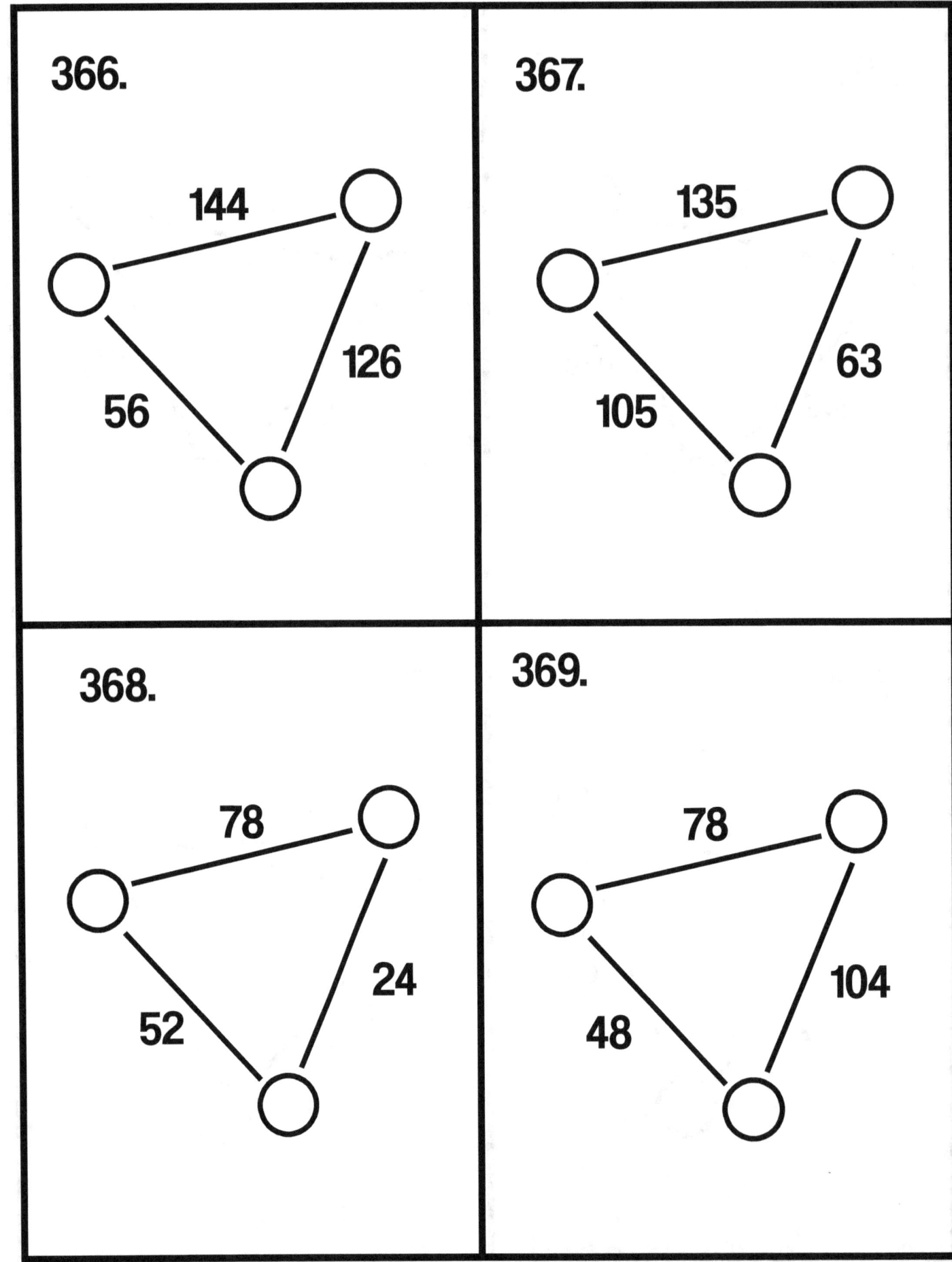

366.

144

126

56

367.

135

63

105

368.

78

24

52

369.

78

104

48

# Juggle 3 - Multiply

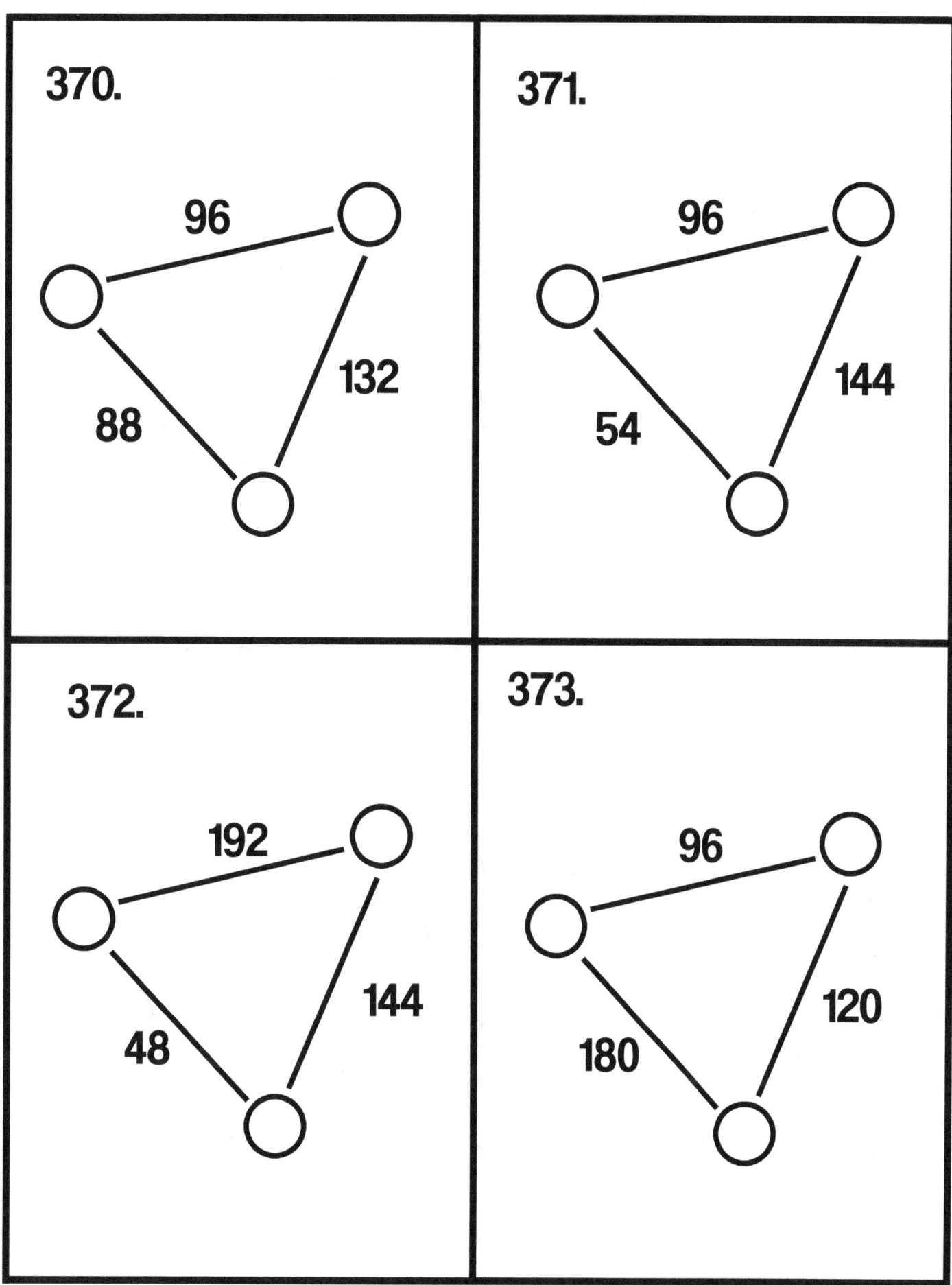

370.

96

132

88

371.

96

144

54

372.

192

144

48

373.

96

120

180

# Juggle 3 - Multiply

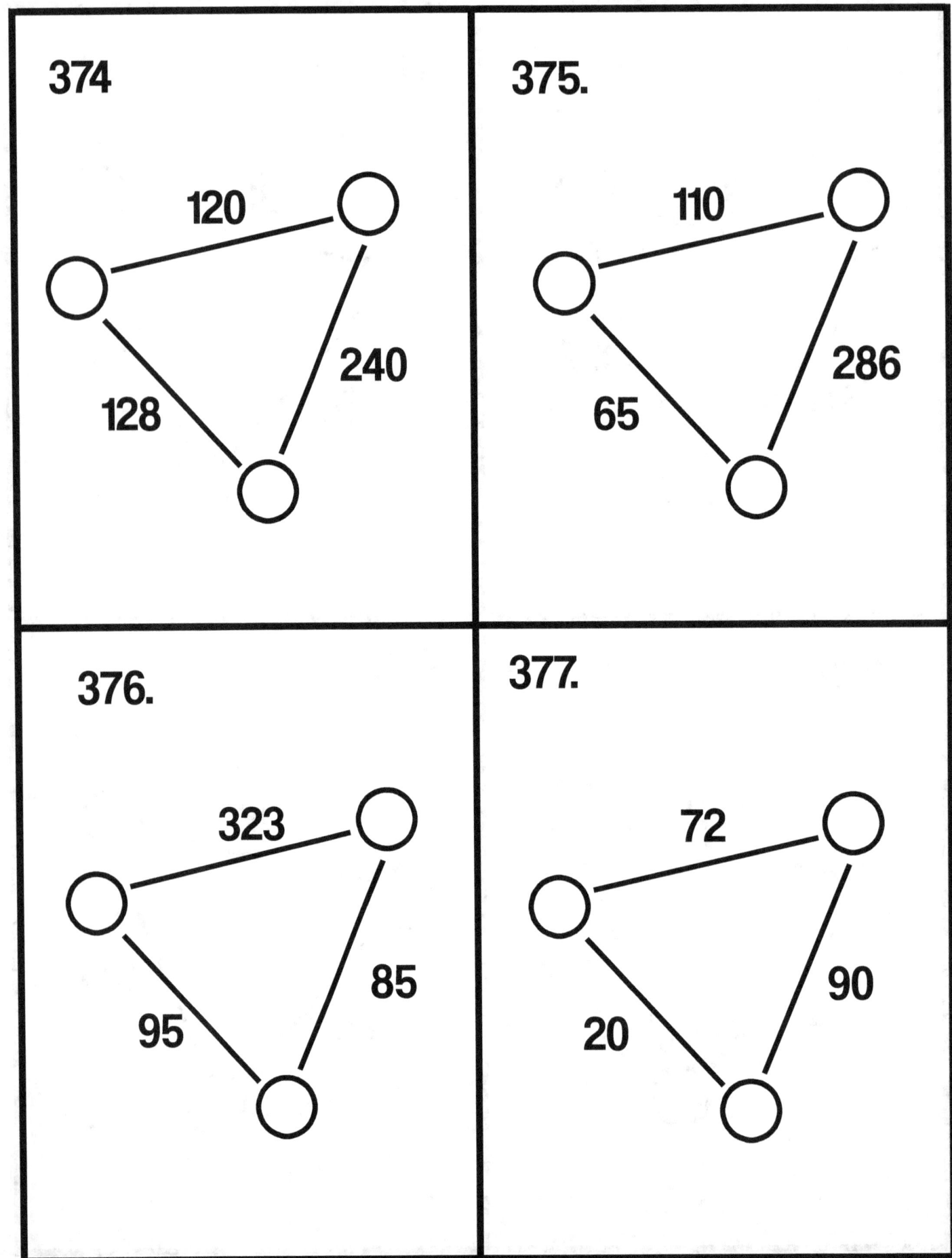

374

120

240

128

375.

110

286

65

376.

323

85

95

377.

72

90

20

# Juggle 3 - Multiply

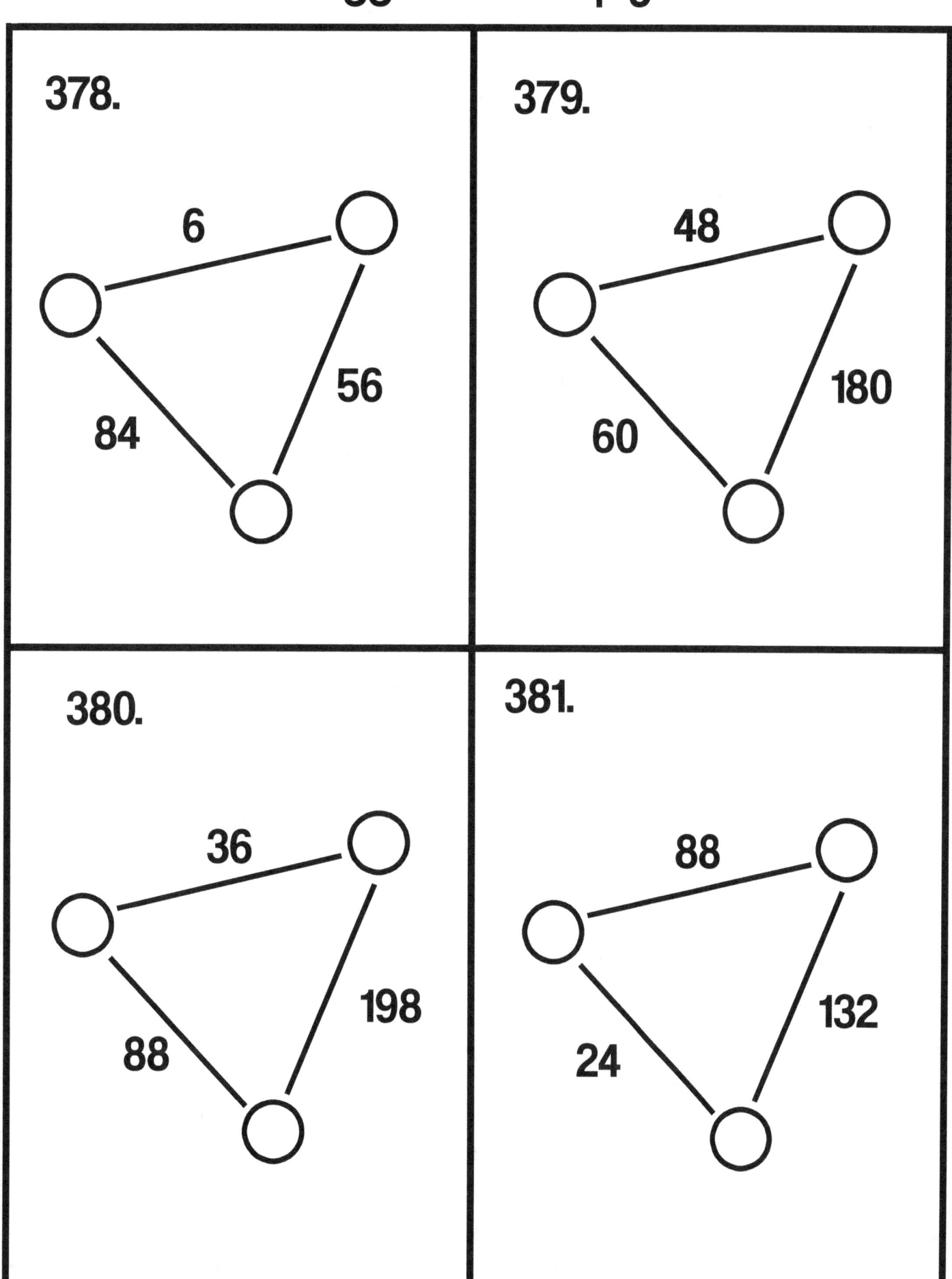

378.

6

56

84

379.

48

180

60

380.

36

198

88

381.

88

132

24

# Juggle 3 - Multiply

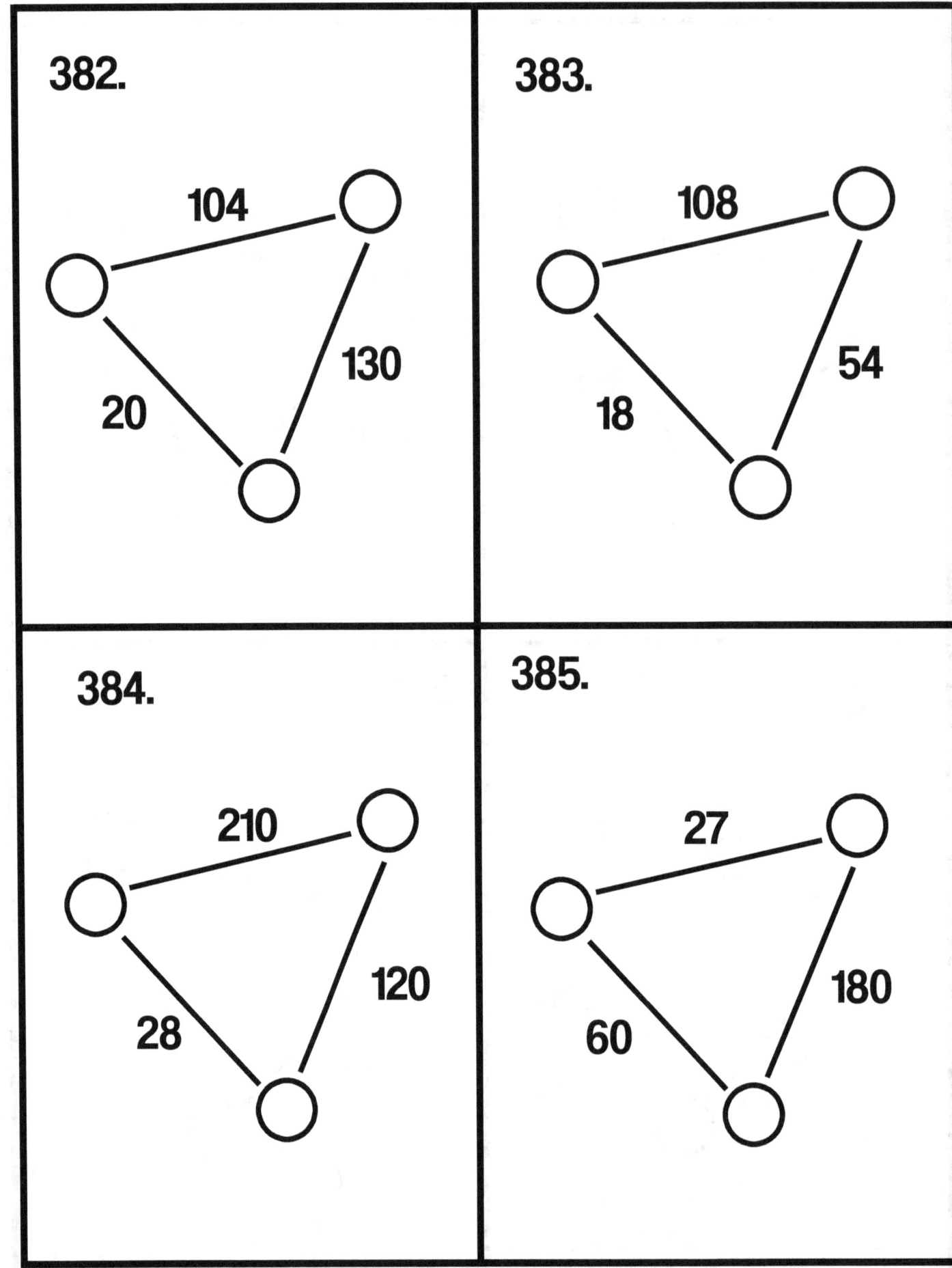

382.

104

130

20

383.

108

54

18

384.

210

120

28

385.

27

180

60

# Juggle 3 - Multiply

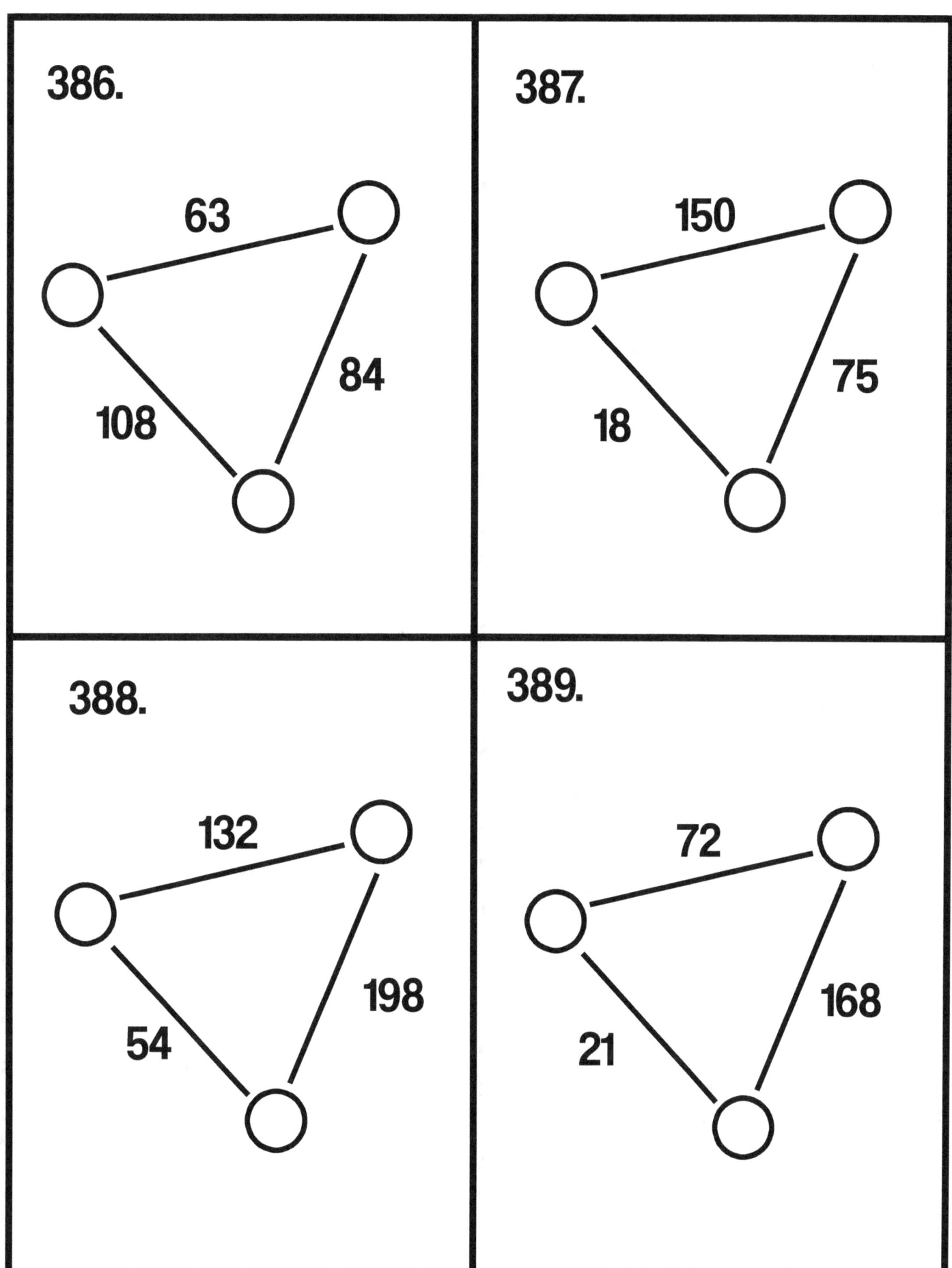

386.

63

84

108

387.

150

75

18

388.

132

198

54

389.

72

168

21

# Juggle 3 - Multiply

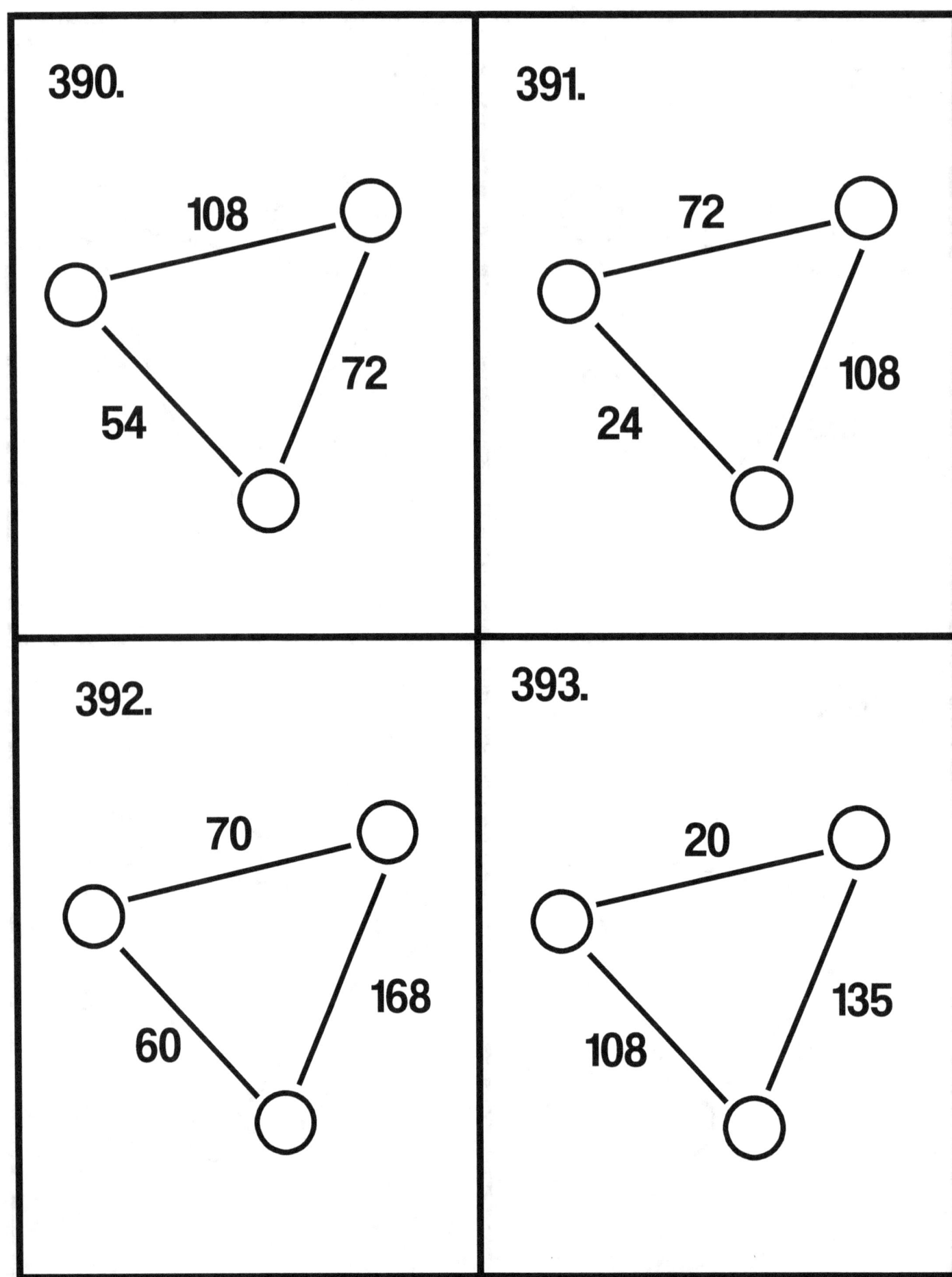

**390.**
108
72
54

**391.**
72
108
24

**392.**
70
168
60

**393.**
20
135
108

# Juggle 3 - Multiply

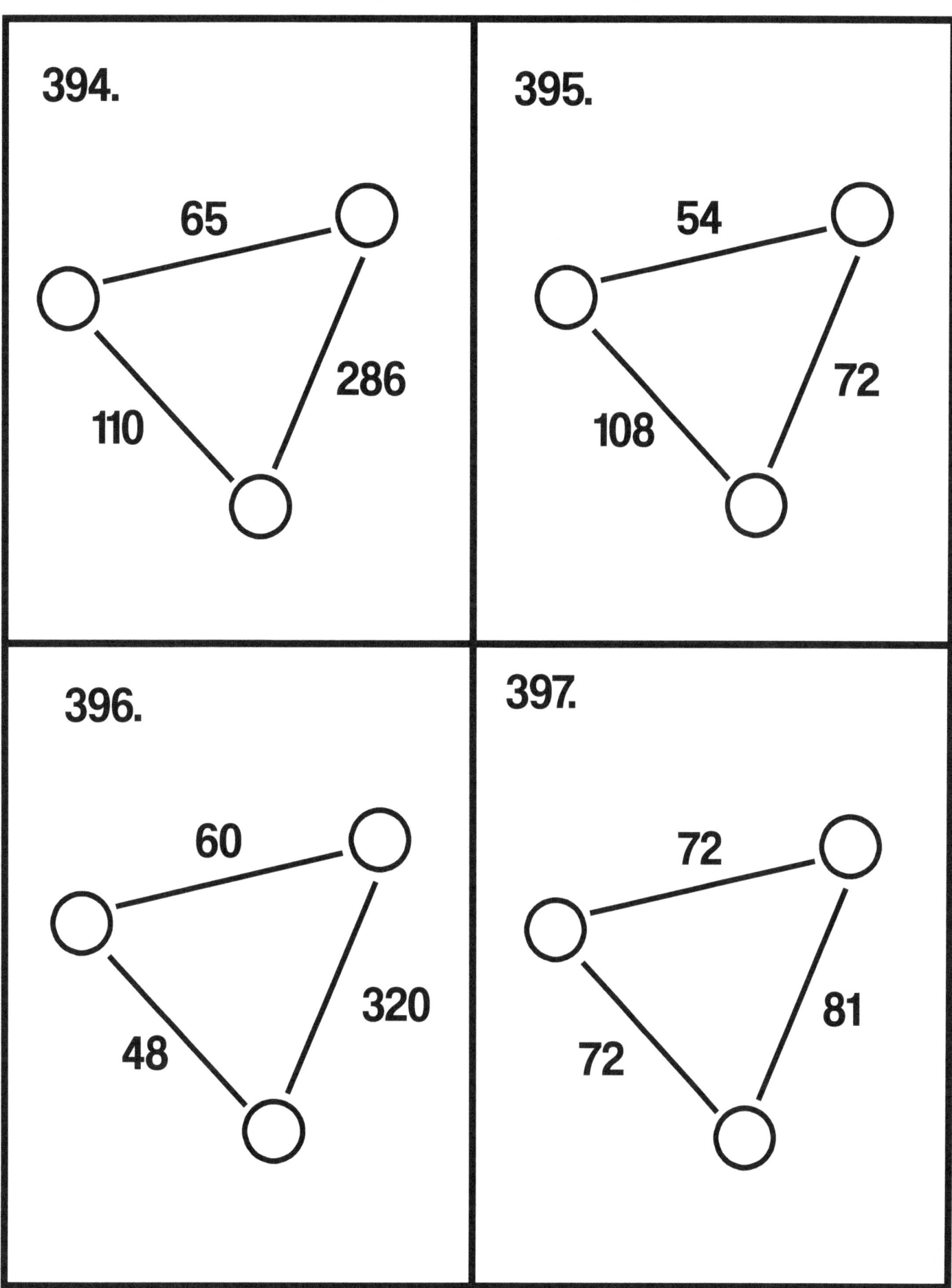

# Juggle 3 - Multiply

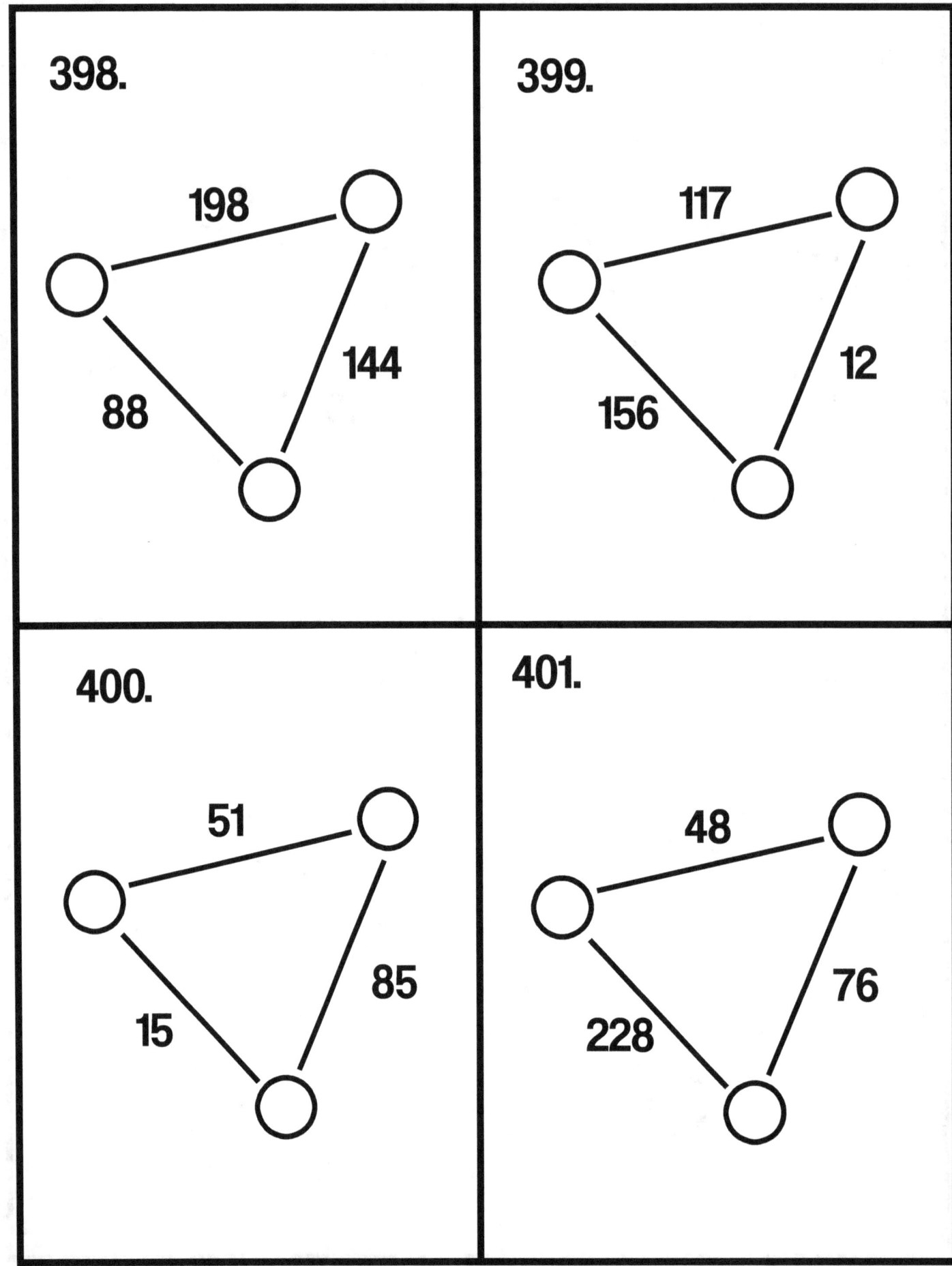

398.

198

144

88

399.

117

12

156

400.

51

85

15

401.

48

76

228

# Juggle 3 - Multiply

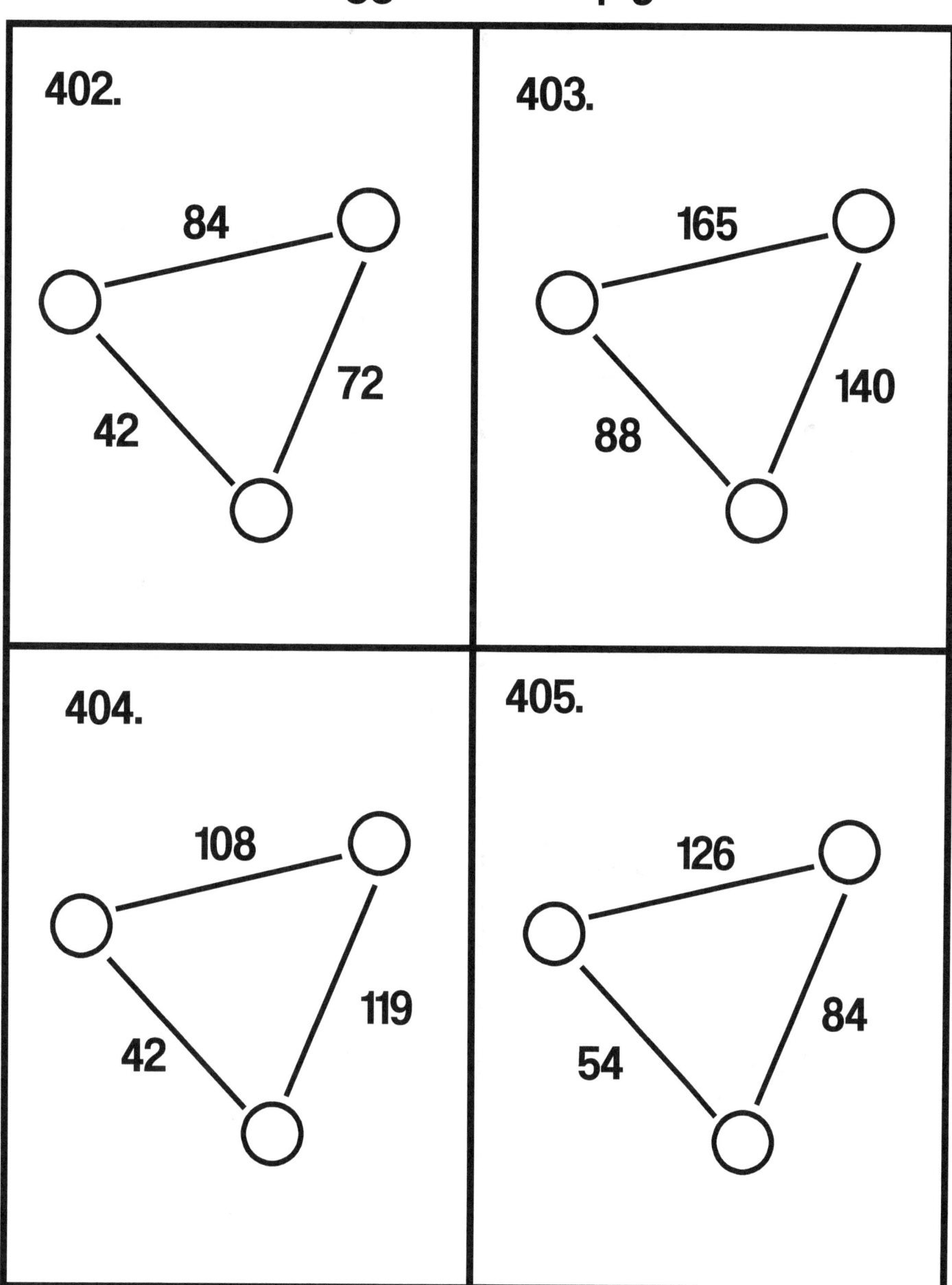

402.

84

72

42

403.

165

140

88

404.

108

119

42

405.

126

84

54

# Juggle 3 - Multiply

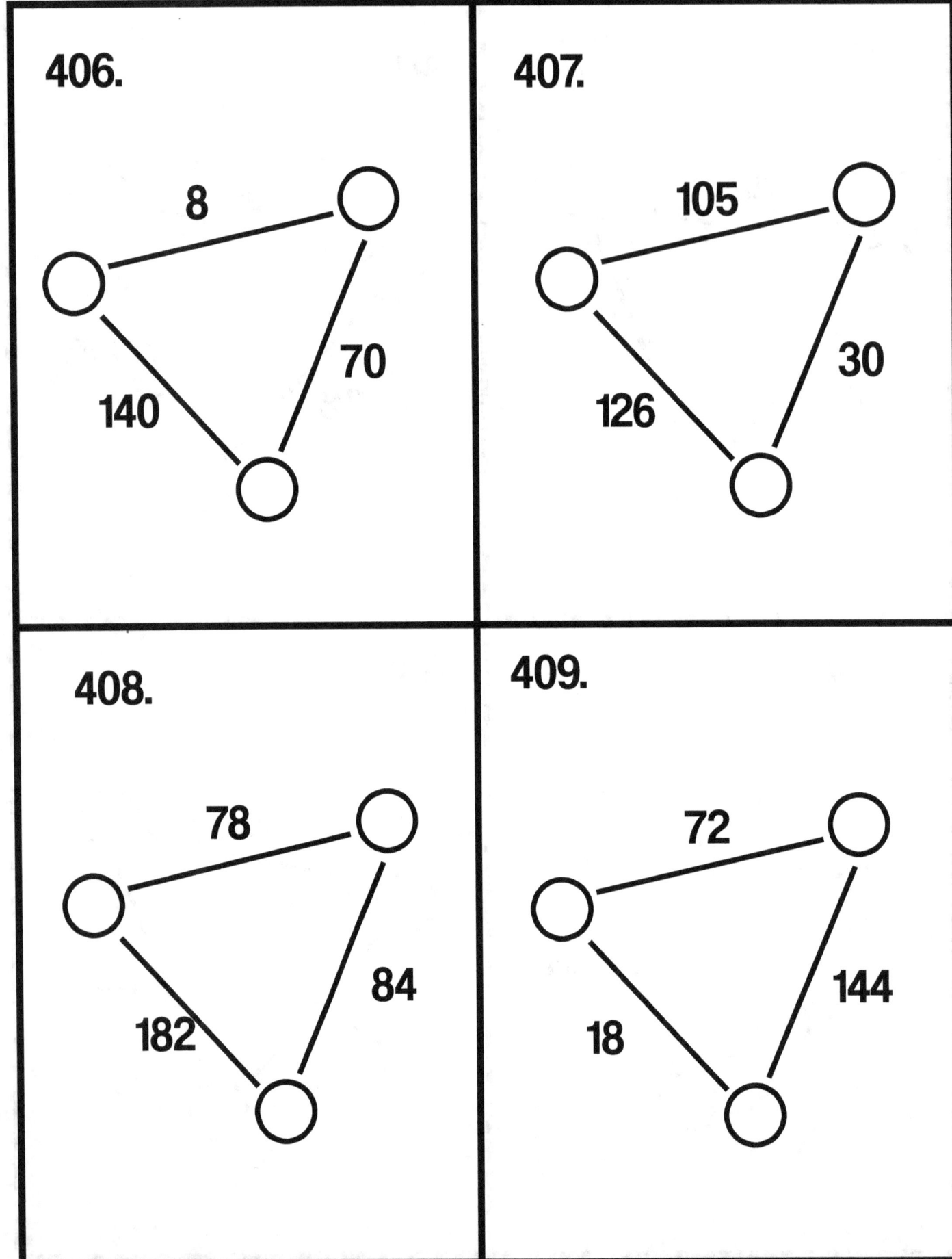

406.

8   70   140

407.

105   30   126

408.

78   84   182

409.

72   144   18

www.ingramcontent.com/pod-product-compliance
Lightning Source LLC
Chambersburg PA
CBHW081138170526
45165CB00008B/2724